Wordsworth Donisthorpe

A System of Measures of Length, Area, Bulk, Weight, Value, Force, etc.

Wordsworth Donisthorpe

A System of Measures of Length, Area, Bulk, Weight, Value, Force, etc.

ISBN/EAN: 9783744743723

Printed in Europe, USA, Canada, Australia, Japan

Cover: Foto ©berggeist007 / pixelio.de

More available books at **www.hansebooks.com**

A

SYSTEM OF MEASURES

OF

LENGTH, AREA, BULK, WEIGHT,

VALUE, FORCE, &c.

BY

WORDSWORTH DONISTHORPE
AUTHOR OF
'INDIVIDUALISM' 'PRINCIPLES OF PLUTOLOGY' 'LAW IN A FREE COUNTRY' ETC.

Printed by
SPOTTISWOODE & CO., NEW-STREET SQUARE, LONDON
1895

PREFACE

EVER since the middle of the present century a strong feeling has prevailed of the immense importance, if not of the absolute necessity, of introducing into England a decimal system of measures of length, area, bulk, weight, and value in place of the tangled and inconsistent medley which now flourishes. Royal Commissions, one after the other, have pronounced emphatically in favour of such a uniform decimal system, and more especially of the metric system used by the French. These reports bear the honoured names of Sabine, Herschel, Airy, and many others. And yet nothing has been done beyond making the use of the metric system permissive in contracts. 'We are of opinion,' runs the Report of the 1868 Commissioners, 'that it is inexpedient that any legislation should take place with respect to the metric system until the whole subject of the weights and measures of this kingdom be brought before Parliament in one Bill.' Good; but successive Governments have been invariably debarred from carrying out these reiterated recommendations by the fact that no single and complete system has ever yet been laid before Parliament. Various schemes dealing with the coinage or with weights, or with other measures have, it is true, been submitted, but nothing approaching to a complete and systematic whole covering the whole field of measurement.

For example, the pound-and-mil scheme was recommended by the

Select Committee of the House of Commons on Decimal Coinage in 1853. The avoirdupois pound has also been put forward as the best unit for a decimal system of weight. Again, for length and area measures the foot has been preferred to the yard as the unit. And so on. Nothing need be said against any of these suggestions beyond the fact that if any one of them is right the others must be wrong, because there is no clear, simple, and finite relation between the foot and the pound avoirdupois, or between the pound avoirdupois and the pound sterling, or between the pound sterling and the foot. It is generally admitted that piecemeal legislation would do more harm than good. Our present chaos of systems is now at its worst. Slight improvements in any branch of it would only have the effect of strengthening the opposition to reform. For instance, if the pound-and-mil system of coinage were in full working order, the worshippers of the 'fixed star'—our wonderful pound sterling of 123·27448 grains of gold $\frac{11}{12}$ fine—would have more to urge in favour of their fetish than they now have. Let us, then, bring forward a complete system of measures before half-hearted reformers have improved their defences.

It is commonly asserted that the English nation is incapable of thoroughness, that it loves compromises and half-measures, that it hates general principles. Surely this is a mistake. When once the doctrine of Free Trade came to be understood, what other nation carried it out so logically and thoroughly as England? When the immorality of slavery received recognition, what nation was so quick or so generously earnest in eradicating the abuse? Was not England at one time alone clinging to the single monetary standard? Where is the principle of free speech and a free Press more consistently adhered to than in this so-called country of compromises? It would almost seem that England, so far from being the least likely, is of all countries the most likely to carry out a great reform in the face of every obstacle, when once the grand principle upon which it is based has become generally recognised. Reverting to the question under discussion, it is probable that England's backwardness in the matter of rational measures is due, not to her inability to realise the advantages and beauties of the metric system, but to her ability to appreciate its real and manifold defects. Let us remove these, and we may safely predict that England will be quick to adopt a system based on principles wider and deeper than those which underlie the systems of any other country. Why legislate in haste and repent at leisure?

It is in the hope of presenting the public with such a complete system of measures that I have undertaken the present work, the labour involved in which cannot fairly be measured by its bulk. When it comes to be seen that this system is complete, that it covers the whole field of measures, that it is suited for an international system, and that it is of all systems the easiest to comprehend and the simplest to work, I have little doubt of its being carried into law with as little delay as possible.

Perhaps I should offer some apology for undertaking what, at first sight, appears to be the business of the Government itself; but it cannot be too often reiterated that, whatever be the merits of democratic government—and they are great and manifold—there are certain tasks which it is utterly incompetent to perform. The very house in which Parliament dwells could hardly have been constructed by our legislators themselves. Suppose a picture to be painted by all our Royal Academicians, each putting in a little bit of his own workmanship, it may be doubted whether, as a work of art, it would command a high price in the market, though possibly as a curiosity—a shocking example of the results of committee work and a fine illustration of the ancient proverb about too many cooks—it would possess considerable value. Similarly, no system of measures produced piecemeal and on the patchwork system by various Royal Commissions (one on currency and monetary standards, another on weights, a third on land measures, and a fourth on gas-meters) can in the nature of things be other than inferior to the production of a single mind.

It would probably be wiser for the present to leave things as they are than to adopt the recommendations of the 1853 Commission, and to decimalise our measures of length, weight, and value from the units yard, pound avoirdupois, and pound sterling respectively. These units are unco-ordinated, arbitrary, and meaningless. My objects in the present work are, first, to show the defects of the existing English arrangement—a not very difficult task; secondly, to show the defects of the French system, for it is far from faultless; and thirdly, to propose a complete system embodying all the merits of the metric system without any of its drawbacks. In addition to this, an attempt has been made to restore the ancient Gothic system —a system which was in many ways superior even to the metric, and in most respects admirable, but the *débris* of which is no more fitted to modern requirements than the lovely ruins of Bolton Priory are suited to the domestic needs of the nineteenth-century householder.

CONTENTS

CHAPTER I

THE DUTY OF THE STATE

Contracts—Definition of Terms—Certain Timid Objections—Negative Compulsion—Individualism and Democracy 1

CHAPTER II

METHOD OF METROLOGY

From Known to Unknown—Inverse Historic Method—Restoration of Original System—Retrace Steps historically—Apply Principles discovered—Build up the System of the Future 9

CHAPTER III

MODERN ENGLISH STANDARDS

The Yard—The Seconds-pendulum—The Fire and the Attempt to reconstruct—Failure—The New Yard—The Pound—Strange Relation between the two—Great Strides in Precision—Nothing else to recommend our Standards—Binary Scale—Origin of some Measures—Consensus of Opinion as to Advantages of Decimal Scale . 12

CHAPTER IV

MODERN FRENCH SYSTEM

Its simple Co-ordination—Its abominable Nomenclature—Useless and mischievous—Difficult to abbreviate—Universal Language quite Utopian at present—Units . 23

CHAPTER V

A SURVEY OF ENGLISH MEASURES

Yard, Nail, Foot, Fathom, Mile, Acre, Furlong—Roman Land Measures: Jugerum, Actus, Pertica—Our Rood—Plan of English Square Mile—Bulk Measures—The Pipe—Liquid Measures, mostly Dutch Names—Original Capacity—Causes of Variations—Some other Bulk Measures—Weights—Five Systems—Battle between the Troy and Avoirdupois Systems—Alliance between Trojans, Physicians, and the Mint—The old Pynd—The Dram—Celtic Ring-money—The Avoirdupois Scale examined—Wool-weight—The Clove—The Wey—The Pack—Sack and Hundredweight—Stone, Tod, and Ton 28

CHAPTER VI

THE OLD PIPE SCALE

An Early Manuscript—Pipe and Peck—Series of Cylinders—Pipe was a Yard Cylinder—Binary Progression—Pipe and Culeus—Gothic Pynd and Kan—The Medium—Wheat—Specific Gravity ·8—The Yard and the Pynd—Objections to the Ratio 62 lb. to the Bushel—Saxon Pennies—The Colonia Pound—Relation to Pynd—Edward I.'s Trew-pound—Names of Double Cylinders and Half-cylinders—Mark and Ounce—Gothic and Roman Scales of Subdivision—Early Gothic Coins—Old and New Grain—Avoirdupois Pound and the Pynd—Henry VIII. invents our modern Troy Pound in 1527—His Blunder—Origin of Colonia Pound—Gold Medium 61

CHAPTER VII

BEFORE THE ACT OF 1824

The Gallons—Pipes, Butts, and Vats—Immense Variety of Liquid and Grain Measures—Over a Dozen lawful Bushels—There were at least sixteen different Stones—Special Measures for various Commodities 82

CHAPTER VIII

GREEK MEASURES

Absurd Ratio between Æginetan, Euboic, and Roman Bulk Measures—Bushel and Peck—Medimnus—The Amphoreus and the Quadrantal—Chous and Congius—Greek Weights—The Talents—Solonian Talent and Mina—Gold Medium—Standard Silver Medium—Solonian Mina a Knathos of Standard Silver—Superiority of Metal Weights over Water or Wheat Weight—Babylonian Kikkar and Maneh—Greek Small Measures—From the Peck to the Egg-spoon—Confusion caused by the Conquest of Greece—Old Greek Pipe Scale—The Greek Grain . 90

CHAPTER IX
ROMAN MEASURES

Ante-Sillian and Post-Sillian—Origin of the Quadrantal—Substitution of Foot for Cubit—Old Sextarius and Modius preserved—New Amphora and Congius—New Ligula or Cubic Inch—Libra and Pondus—The Sicilian Litra—Curious Relation of Litra and Talent—Accidental—Roman Standard Silver—Money Pound and Market Pound 102

CHAPTER X
FROM THE EARLIEST TIMES TO THE FRENCH REVOLUTION

Aryan Scale—Yard and Cubit—The original Cubit a Thousandth Part of Distance done by Surface of Earth in one Second at the Equator—Homologous Greek, Roman, and Gothic Measures—Meaning of Euboic—Euboic Ratio was Ten-twelfths, but Roman and Gothic Ratio was Eight-tenths—Reason for this Difference—The great Roman Metric Reform—Its Ingenuity—The Formula by which the Peck and Pound were kept intact—Origin of Duodecimal Scale—Decay of the Pipe Scale—French Revolution and Birth of the Metric System 108

CHAPTER XI
HISTORY OF THE METRIC SYSTEM

First Step—Vain Appeal to England—Talleyrand's Report—The Metre—The Academy's Report—The Law of April 1795—New Measures for Old—Various Changes in the Nomenclature—Napoleon tries Optional System—Metric System compulsory and exclusive from First Day of 1840—Progress in other Countries—German Zollpfund—Half-steps—Report of the Federal Parliament—Progress in America slow—Adopted by most other States—England cautious and slow—Mr. Chisholm's Objections met 113

CHAPTER XII
THE SYSTEM OF THE FUTURE

Nomenclature—Decimal Scale—The best Unit—Co-ordination—The Basic Arrangement—A single Standard, but no Measure Units—Great Importance of Abbreviation 130

CHAPTER XIII
THE BASIC SYSTEM

Suggested Nomenclature—Measures of Account—Alphabetical Relation—Fractional Measures, if required, unobjectionable—Easy to learn—Hard to forget—Length, Area, Bulk, Weight—Large Units much required—Force, Power, Pressure, Temperature and Heat, Specific Gravity—Value 141

CHAPTER XIV
MONEY

Relative Positions of Money and Credit—Tokens—Legal Tender—Standards of Value—State Coining and State Swindling—Simple Relation of Value and Coin-weight—Cost of Minting—Guarantee against Loss by Wear—International Currency—A true Silver Currency—Multiple Currency—A Right to an Iron Currency—Analyses of Cost of Minting true Money—Silver Coinage expensive—'What is a Pound?'—No State Risk—Coin-insurance should be put out to Public Competition . . 162

CHAPTER XV
SUMMARY . . 189

APPENDICES

I. Table of Metric Weights and Measures taken from the Law now operative—Decree of August 1, 1793—Law of April 7, 1795—Vacillating Nomenclature of the French System 195

II. Index of English Measures in 1800 204

III. A Bill to amend the Law relating to Measures 222
  Schedule I.—Part I. will describe the Imperial Standard Mete and Yasp constructed under the direction of H.M. Treasury; and Part II. the Parliamentary copies of the same.
  Schedule II.—List of Useful Measures.
  Schedule III.—Old Measures in Terms of New, and New Measures in Terms of Old.

PLATES

SOME UNITS OF THE BASIC SYSTEM AND MEASURE OF ACCOUNT IN THE BASIC SYSTEM	*Frontispiece*
JUGERUM	*To face p.* 34
SQUARE MILE, SQUARE FURLONG, ACRE, ROOD, AND PERCH . .	,, 38
THE OLD PIPE SCALE	,, 70
OLD ENGLISH SILVER WEIGHT	,, 74
THE FIVE POUNDS AND THEIR SUB-DIVISIONS . . .	,, 78
GREEK BULK MEASURES AND WEIGHTS	,, 100
SEXTARIUS	,, 102
GREEK AND OLD ROMAN PIPE SCALE . . .	,, 106
PIPE SCALE	,, 112
THE FRENCH MEASURES TABULATED ON THE BASIC SYSTEM . . .	,, 116
FOIL, SEAL, AND EN, FULL SIZE, AND RELATIVE SIZES AND SHAPES OF THE ACRE, WORTH, AND JUGERUM	,, 148
DESCRIPTION OF COINS, AND COINS IN ORDER OF DIAMETER . .	,, 160
SPECIMEN OF ACCOUNT BOOK	,, 188
ARYAN PIPE SCALE	,, 192

MEASURES OF ACCOUNT IN THE BASIC SYSTEM

Length	Area	Bulk	Weight
Jot	En	Ove	Une
Quil	Seal	Die	Gram
Hand	Foil	Litre	Yasp
Mete	Nap	Vat	Ton
Beam	Ar	Keep	Poid
Course	Worth		
Reach	Ing		

Directions

No figures are required. The length measures rise by tens, the area measures by hundreds, and the bulk and weight measures by thousands. Thus :—

    10 jots make a quil; 10 quils make a hand; and so on.
    100 ens make a seal; 100 seals make a foil; and so on.
    1,000 oves make a die; 1,000 dies make a litre; and so on.
    1,000 unes make a gram; 1,000 grams make a yasp; and so on.

The area measures are the squares of the length measures on the same line. The bulk measures are the cubes of the length measures on the same line. The weight measures are the weights of pure water at freezing-point of the bulk measures on the same line.

SOME UNITS OF THE BASIC SYSTEM

Unit of time is one second: symbol z.

Unit of value is one gram of pure gold; symbol X.

Unit of specific gravity is that of pure water at 0ᶜ (freezing): symbol $\frac{t}{v}$ for solids and liquids, $\frac{y}{v}$ for gases.

Unit of force is the mete-yasp: symbol $m \cdot y$.

Unit of power[1] is 100 mete-yasps per second, called a *power*: symbol $\frac{c \cdot y}{z}$.

Unit of pressure is one yasp per seal, called an *atmosphere*: symbol $\frac{y}{s}$ or $\frac{p}{a}$.

Unit of temperature is that which is added to any weight of water by a fall of one course: symbol 1ᶜ.

Unit of heat is the heat produced by the fall of one yasp of water through one course: symbol $c \cdot y$.

---

[1] This is not the same unit selected in Chapter XIII.; but being about 1½ of one present horse-power, it will be admitted, I think, that it is a more suitable unit of power than a quarter of a horse-power. Of course all these units are open to question. Not being an autocrat I am unable to thrust them on public acceptance by force, and not being a socialist I am unwilling to do so

A SYSTEM OF MEASURES

CHAPTER I

THE DUTY OF THE STATE

IF there were no State, individuals would necessarily form voluntary associations for doing many things now done by the State—the warding off of foreign aggression, the stamping out of criminal aggression at home, the settling of disputes between fellow-citizens on general principles, and perhaps some others. I shall assume that the duties I have named properly appertain to the State. Some individualists will contend that the State must always perform these functions; anarchists will admit that in the present state of social development (whatever the future may have in store for us) the State *has* to perform these functions, and that, so long as this is the case, it should perform them well.

Leaving on one side foreign affairs and criminal law, let us look at the remaining business of settling disputes between citizens. Take the case of an agreement between two parties. The State is in precisely the position of a pre-appointed arbitrator, or of an umpire in a game. Instead of pre-appointing one individual, the contracting parties agree to appoint all their fellow-citizens. These fellow-citizens appoint a sub-committee or a judicial body. Surely no one will be bold enough to deny that this judicial body very rightly lays down certain conditions. It will not undertake to decide every dispute whatever, but only such as can be fairly settled without an unnecessary waste of public time and energy. For example, it says (and a private arbitrator would rightly say the same), 'Every agreement to buy and sell land must be made in writing.' This condition is reasonable, because it saves endless trouble in sifting a mass of conflicting evidence in the absence of a written agreement, with a very slight chance of arriving at a just decision. The State refuses to listen to certain classes of evidence on the alleged ground that it is irrelevant, but really and truly because the sifting and weighing of it is a waste of time. The State enjoins the use in some cases of certain particular words (such as the word 'limited' in a class of joint stock undertakings), because it is very easy to use the word, and very difficult to decide some questions of liability when the word is not used. All this is quite reasonable, and exactly what a private preappointed arbitrator might wisely insist upon. No prudent man would undertake to decide every case on its merits. He would generalise, classify, and define injuries, just as the State does now. Promises must needs be made in language—words. The arbitrator has to interpret these words; surely he may stipulate as to what language he will undertake to interpret. There is no coercion here. All the arbitrator does, if other words are used, is to say, 'I cannot undertake to interpret these words; you must settle your dispute outside.'

Suppose that two men agree the one to buy and the other to sell a parcel of land measuring 35 hogsheads and 27 scruples, and suppose the agreed price is 75 minas and 3 ducats per hogshead; I see no valid reason for interfering with them. But if they quarrel over the meaning of the terms, and come to the court to settle their differences, I see a very strong reason for not wasting the public time in trying to find out exactly what they meant by the words of the agreement. I say it is no interference with private liberty for the State to declare beforehand what terms it will undertake to interpret. And, descending from the general to the particular, the definition of measures of length, surface, bulk, weight, and value seems to me as properly a State function as the definition of crimes and of injuries. Clearly, if the

State declined to define the various classes of offences, there could be no such thing as a body of criminal law. Every case would have to be decided on its merits, and chaos would supervene. Certain individualists, with whose principles I am in general accord, say, 'Let people buy and sell in what measures they choose; the State has nothing to do with it, except to interpret their agreements and to enforce them.' This is, I hold, a mistaken view. Most people agree with me, and the State itself agrees with me—two very strong reasons against me, I admit. The State has at one time or another defined nearly all our measures by Act of Parliament, and to-day it defines the yard, the pound weight, the gallon, and the pound sterling. If the Government were to say, 'We will have nothing to do with weights and measures; let people manage their own affairs,' one could forgive them. But from the days of Edward the Confessor, or earlier, down to the present reign the State has regularly, every few years, asserted its claim to regulate all our measures by Act of Parliament.

Here, I think, we shall find ourselves on common ground; whether the State ought, or ought not, to undertake the regulation of measures, if it actually does undertake it, then, in that case, it is bound to do it as well as possible. It cannot in one breath declare, on the one hand, that this is a State function, and on the other, that the State has nothing to do but to follow the lead of custom. And yet, in the very Report in which the introduction of the new imperial gallon is recommended in 1819, we have the following inconsequential observation :—

'A general uniformity of weights and measures is so obviously desirable in every commercial country, in order to the saving of time, the preventing of mistakes, and the avoiding of litigation, that its establishment has been a fundamental principle in the English constitution from time immemorial, and it has occasionally been enforced by penal statutes, and by various other legislative enactments. At the same time, it has commonly been considered as one of those objects which cannot, consistent with logical accuracy, with natural justice, and with the liberty of the subject, be very precisely defined, or very peremptorily and arbitrarily enjoined, on every occasion; and there are many instances in which a departure from complete uniformity is not only tolerated, but established by law. It must, indeed, sometimes be almost as impossible to control the despotic influence of custom with respect to the contents of a measure of a certain denomination as with respect to the signification of a word of any other nature; and even the terms of number, precise as they necessarily are in their strict meaning, have become liable to perpetual variations, according to the objects to which they are applied. And these variations, however inconvenient they may appear upon a general view of the subject, have been repeatedly sanctioned by their adoption in the acts of the legislature.'

Again, in the 1869 Report, which recommends the introduction of the metric system into England, we are told that at the present time there is no evidence to show that any considerable portion of traders and their customers in this country are dissatisfied with the imperial system now in use, or that they desire to substitute the metric system for it. I am not aware that the Gold Coast negroes have expressed any dissatisfaction with their cowrie coinage. 'It is obvious,' say the Commissioners, 'that in this country, where the people are more accustomed to self-government than in other European countries, the executive has far less power of compelling obedience to the law in all the small transactions of trade against the wishes of the public. Should an attempt be made at the present time to introduce the metric system by legal compulsion, the Commission regard it as certain that very great confusion would be produced, and they think it highly probable that the attempt would be met by such an amount of resistance, active and passive, that it would totally fail.' The whole force of this argument turns on the meaning of the word compulsion. The Statute of Frauds is not usually regarded as a coercion Act; it is merely an Act defining what species of agreement the State will enforce, and what species it will not. On all fours with the Statute of Frauds in this respect is the recommendation in the Report of 1841: 'That every contract be declared void, in which the quantity of the matter or thing of which the contract treats shall be expressed in denominations different from those proposed to be legalised, or shall be expressed in any customary weight or measure of any place different therefrom, or according to any usage or mutual understanding between the parties to the said contract; unless, in the case of such contract, the denominations, customary weights or measures, usages or mutual understandings be expressly defined and explained in the body of the contract, and in the operative clauses thereof, in terms of the said legalised denomination.' This simple and non-coercive provision was probably based on the analogous section of the loi du 4 Juillet 1837 : ' Il est défendu aux juges et arbitres de rendre aucun jugement ou décision en faveur des particuliers sur des actes, registres ou écrits dans lesquels les dénominations interdites par l'article précédent auraient été insérées, avant que les amendes encourues aux termes dudit article aient été payées.' Even more drastic is Section 15 of the Act of 1824: 'All contracts, bargains, sales, and dealings, for any work to be done or goods to be sold by weight or measure, where no special agreement shall be made to the contrary, shall be deemed, taken, and construed to be made according to the standard weights and measures ascertained by this Act. And in all cases where any special agreement shall be made with reference to any weight or measure established by local custom, the ratio or proportion which every such local weight or measure shall bear to any of the said standard weights or measures shall be expressed, declared, and specified in such agreement; or, otherwise, such agreement shall be null and void.'

Section 19 of the Act of 1878, which is now operative in this country, is as follows: 'Every contract, bargain, sale, or dealing, made or had in the United Kingdom, for any work, goods, wares or merchandise or other thing which has been or is to be done, sold, delivered, carried, or agreed for, by weight or measure, shall be deemed to be made and had according to one of the imperial weights or measures ascertained by this Act, or to some multiple or part thereof, and if not so made or had shall be void.' Not content with this non-coercive provision, the Act proceeds to apply coercion: ' Any person who sells by any denomination of weight or measure other than one of the imperial weights or measures, or some multiple or part thereof, shall be liable to a fine not exceeding forty shillings for every such sale.' This is not only coercion, but quite unnecessary coercion. There is no conceivable reason why men should not exchange their goods in any quantities and in any measures they choose, provided they do not ask the State to understand and interpret their arbitrary terms. I trust I have made it clear that I propose nothing which can fairly be described as compulsion, or as State interference with the liberties of the citizens. On the contrary, I propose to leave the public in the possession of full power to manage their own affairs in their own way, and, so far from adding to the powers of the State, I propose to deprive it of the function of interpreting words which are foreign to its legal vocabulary.

'Let things slide,' is the motto of the stick-in-the-mud party. It was chiefly in this sense that Lord Melbourne used to say, 'Can't you let things alone?' But the individualist of to-day is a very different person. He does not say, 'Let that which is filthy be filthy still.' He says to the State, 'Hands off! You are not the proper party to meddle. You always make matters worse, and we want to see things made better.' No wonder there is such a strong and bitter feeling throughout all ranks of society against individualism, when able men regard it as the principle of All-right. Things are not All right. Individualists know that. They are not exclusively the comfortable party. They desire reform, and very considerable reform. But they want reform on the right lines. Does any sane person really believe that the army and navy will ever be brought into proper condition through the force of public opinion? Is it, then, in his eyes, necessary to satisfy the newly enfranchised of the suitability of the Scilly Isles as a naval station? Are we to get up an agitation to supplant the regulation rifle by something which the rabble consider better? The next thing will be for the Trustees of the National Gallery to invite the Trafalgar-square ground-floorers to step inside and decide as to the hanging of the pictures. It is precisely this ridiculous assumption of power on the part of the people, through their amateur agents, which has brought the services to their present state of supreme inefficiency. Nothing else. By all means let the people keep tight hold of the purse-strings. Let the executive in all departments be selected on sound principles. But, having chosen the best men available, let us trust them. Unless we can trust the Administration to declare war, for instance, without the previous consent of Parliament, we are handicapped in our diplomacy with respect to other Powers, just as a whist-player would be handicapped who put his cards on the table while his opponents kept theirs in their hands. It is the same in every department. We do not want the public opinion on every detail of government. We want the public to choose its servants, and to trust them to do what is best. Experts must know more about all matters than even the inspired 'free and independent elector.'

What has the State done in its amateur capacity? Among other things, it has undertaken the letter-carrying monopoly; it has undertaken the money-minting monopoly; and it has undertaken the regulation of weights and measures—these in addition to its normal functions of defending citizens against foreign and internal aggression. And how has it succeeded? I need not add to the almost continuous indictment which has been brought against the Post Office of late. Let us look fairly at the coinage question. It does not require a mathematical genius to perceive that a system of coins bearing the relation to one another of 4 12 20 verges on the idiotic. Even the State acknowledges that. Somebody took the trouble to calculate the time annually consumed in turning farthings into pence, pence into shillings, and shillings into pounds, in English account books. It comes, on the lowest valuation of human labour, to millions of pounds sterling. What did the State say about the matter some forty years ago through the mouth of the Parliamentary Committee of 1853? That while fully concurring with all the witnesses examined as to the great advantages of a decimal system, and while recognising that the change would necessarily be accompanied by some inconvenience, they desired to record their conviction that the obstacles were not of such a nature as to create *any doubt of the expediency* of introducing that system; and that the inconvenience would be *far more than compensated* by the great and permanent benefits which the change would confer upon the public, and to a still greater extent upon future generations. Could anything be plainer? A whole generation has passed away, and absolutely nothing has been done in this direction beyond the passing of a permissive Act and the coining of the florin.

CHAPTER II

METHOD OF METROLOGY

In all departments of inquiry it is well to proceed from the known to the unknown. More especially is this the case in the science of metrology. Otherwise, we are driven to base all our estimates of ancient measures on the crudest guess-work. The Roman libra is said to be a simple fraction of the Æginetan talent, and an approximation is hailed as confirming an inspiration. Then the Greek talent is said to be based on the Babylonian foot. Why? Because the Greek foot is too short to fit in with what is known of the weight, and with the current hypotheses. Not that there is any evidence that the third part of the Babylonian double cubit was longer than the πους; while, on the other hand, there is evidence that it could not be so much longer as to satisfy the requirements of the theory. Let us begin with what is before us. Let us examine our own weights and measures; follow them back to their earlier forms, and so grope our way to the measures used by our predecessors, the Goths and the Romans. It is probable that by following this inverse historic method we shall avoid many foolish blunders, otherwise almost inevitable.

Having established certain fixed points from which we can reason without doubt or danger, we shall presently be in a position to establish new centres, to increase the number of fixed points, and as these multiply, the work will become both easier and surer. At present all is misty, uncertain, and disputed. Even the unit of weight of the great Roman Empire—the libra—is variously computed. Numismatists put it at about 5,040 or 5,050 tr. grs. Metrologists put it at 4,989. Some put it (on the faith of the Tauromenian inscription) as high as 5,540. Those who have measured the Farnese congius make it about 5,200 or less. I shall show that before the establishment of the amphora quadrantal it was 5,051 tr. grs., and after that date 4,989. Without certain established facts we are helpless. We can drift about and make ingenious guesses, and that is all. I propose, therefore, in the present work to begin with a survey of our modern English measures, and more especially of the measures in vogue before the Act of 1824; to trace them all back as far as they can be followed; to ascertain by a process of induction what, in all probability, was the prototype of each of them, and what was the origin and the exact meaning of that prototype. If any convergence is observable I shall follow up the converging lines, and so arrive (if the facts warrant us) at the skeleton of an earlier scheme, and effect what may be called a restoration of the English system; just as from a few foundation stones, the *débris* of an old ruin, architects present us with a restoration of the original building; and just as Cuvier or Owen could give us a life-like picture of an extinct animal from the examination of a few of its bones or teeth. If palæontologists had followed the method hitherto adopted by metrologists, our ideas of the mastodon, the megatherium, and the pterodactyl would be very different from what they are, and very wide of the truth.

Having, so to speak, restored the Gothic system, I shall proceed in like manner to the study of the Roman, and more particularly the Greek system, and if, in the process of excavation, the foundations justify us in concluding that all three were originally based on the same ground plan, and that they were, in fact, merely varieties of one grand type, we shall then be able to argue from one system to the other with profit and with safety. We shall then, and not till then, be in a position to write the history of measures of size and weight from the earliest times down to the present. And in course of the survey we may be sure that certain principles will thrust themselves upon us which may serve as foundations for the System of the Future. On our return journey from ancient to modern times we shall pause

METHOD OF METROLOGY

at the year 1889, and pass in review the history of the French Metric System, from which, also, we shall extract principles of value.

The next step will be to marshal and combine the principles educed from the world's practice, and in this light to criticise, first, the Old World System or Systems, and secondly, the Modern French System. We shall doubtless find merits and defects in each. And finally, with the experience of the past to guide us, we may fairly expect to be more successful in forecasting the system of the future than those who have gone before, and to whose failures in certain respects we are indebted for our present knowledge.

Such in outline is the method of this work. As to the details, it will be necessary to make use of all the resources of modern science, and particularly of etymology, which often throws a light—or, at least, furnishes a clue—when all else fails.

CHAPTER III

MODERN ENGLISH STANDARDS

LET us now look a little carefully into our modern English system of measures, and let us confine ourselves for the present to the measures of length, area, bulk, weight, and value. Our unit of length is the yard—what is a yard? To begin with, it is nothing particular in itself; it is not the length of anything in heaven above or in the earth beneath. What it pretends to be no one knows, but the way to find it is this: take a pendulum and adjust it till it vibrates seconds of mean time in the latitude of London in a vacuum at the sea level. You will find that it is of a certain length—always the same length. We cannot name it because we have not yet any standard of length. Divide it into 391,393 equal parts. Take 360,000 of these parts, and call it a yard. Why not take the whole thing and call that a yard? Or if this, for some mysterious reason, is unsuitable for the standard of length, why not take one-half, or seven-eighths of it? Why, in the name of common-sense, take such a ridiculous and troublesome fraction as $\frac{360000}{391393}$?

Of course the answer is simple. The yard was settled before; it was the length of an old brass rod of a semi-sacred character, which has always been called the standard yard, and which bore no definite known relation to anything whatever, except perhaps Henry I.'s arm. When the two lengths were compared—the old brass rod and the pendulum swinging seconds in London at sea-level in a vacuum—it was found that the former bore to the latter the ratio expressed by the awful fraction aforesaid. In case this precious yard should ever be lost, the Commission of 1819 recommended that some means should be devised of recovering it. Now, had it been the length of the seconds pendulum under the specified conditions, this would have been theoretically an easy matter. Such, however, was not the case. What, then, was to be done? 'We propose that it should be declared, for the purpose of identifying or recovering the length of this standard, in case it should ever be lost or impaired, that the length of a pendulum vibrating seconds of mean solar time in London, on the level of the sea and in a vacuum, is 39·1372 inches of this scale.' What was to be done, therefore, was to take this pendulum, to divide it into 391,372 equal parts, and to cut off 360,000 such parts. This, would of course, be a yard. Then came the Act of 1824, which, after correcting the length of the pendulum to 39·1393—that is to say, adding $\frac{21}{10000}$ of an inch to its length, proceeded thus : ' Be it therefore enacted and declared, that if at any time hereafter the said imperial yard shall be lost, defaced, destroyed, or otherwise injured, it shall and may be restored by making a new standard yard, having the same proportion to such standard aforesaid as the said imperial standard yard bears to such pendulum.' All was now safe, but no one of course anticipated such a painful eventuality. One short decade passed, and the said imperial standard yard *was* lost, defaced, injured, and, in fact, utterly destroyed in the fire of 1834. And now the wise men were gathered together and commanded to restore it in the manner prescribed. But firstly they could not agree about the length of the pendulum, and secondly they could not divide such pendulum—or indeed any pendulum—into 391,393 equal parts. And so the poor yard was lost for ever, and Section 3 of the Act of 1824 is left to us as a monument of State imbecility.

Instead of rejoicing at so happy a release from an ancient fetish, our rulers actually set to work to make another as nearly as possible of the same length, and after twenty years of vain effort, when Commissioners were appointed to superintend the construction of new Parliamentary measures—and not, be it noted, the restoration of the old ones—their very first recommendation was that in future the conservation of the standards should rest on something better than a sheer impossibility. The Report of 1854, just twenty years after the fire, was clear and cautious : ' After due consideration of this question we adhere to the recommendation contained in the Report of 1841, that no reference be made to natural elements for the values represented by the standard.' Quite so. All this tall talk about millionths of an inch, about pendulums and vacuums and tenths of a degree Fahrenheit, and about the barometric pressure at the level of the sea, in the latitude of London, was magnificent in theory and on paper and *before the fire*, but quite another thing in practice, on a bronze bar, *after the fire*. All these questions are deeply interesting, but let the State leave them entirely to speculative philosophers, cheering them on and occasionally paying the bill. Surely this is the meaning of Recommendation 41 : ' We consider the ascertaining of the earth's dimensions and of the length of the seconds pendulum in terms of the standard of length, and of the weight of a certain volume of water in terms of the standard of weight, as philosophical determinations of the highest importance, to the prosecution of which we trust Her Majesty's Government will always give their most liberal assistance ; but we do not urge them on the Government at present as connected with the conservation of standards.' First we have ten years of somewhat bombastic promise ; then the fire, followed by twenty years of abject failure and impotence ; and finally, in 1854, this humble State confession of weakness, coupled with the prudent resolve in future to leave science to the scientists, and to restrict its energies within its own normal field of activity—seeing that citizens get what they pay for.

The new standard is a solid square-section bar of bronze about half as long as a tall man can stretch with his extended arms. Near the ends of this bar, which is about as thick as your thumb, little holes are sunk to the depth of half the bar's thickness. These two little holes are plugged up with gold pins made flush with the surface of the bar, and a point is marked in the middle of each gold circle. The distance between these two points is called an imperial standard yard. The law tells us the precise thickness of the bar and its exact length and the distance between the pins; but as all these measurements are expressed in terms of the bar itself, I do not see that they give us any information whatever.

The State is now in a position to answer at once the question, 'What is a yard?' It is 'the straight line or distance between the the centres of the two gold plugs or pins in the bronze bar, when the bar is at the temperature of 62° Fahrenheit's thermometer, and when it is supported on bronze rollers placed under it in such a manner as best to avoid flexure of the bar, and to facilitate its free expansion and contraction from variations of temperature.' That is a yard. Now we know. This precious bar is deposited in the Standards Department of the Board of Trade in the custody of the Superintendent of Weights and Measures.

But suppose a fire should break out, even at the Board of Trade, and the country should again be plunged into despair—a yardless country—what is to happen then? To provide against this calamity, four copies of the bar have been made. One has been sent to the Royal Mint, another to the Royal Society, a third to Greenwich Observatory, and the fourth has been immured in the new palace at Westminster. It is not likely that fires will break out in all these places at once, and those bars which are destroyed will be reconstructed on the model of those which are left.

Let us be fair. This rather ridiculous bar, though no length in particular, is just as good in itself as any other standard so long as it can be kept the same length. And the precautions taken for this purpose are perhaps as good as can be expected. The main objection to it is that it is not the exact length of the French metre, and the main objection to the metre is that it is not exactly the length of the English yard. If we are to have international uniformity, one of the two must give way, and since the metre has a meaning while the yard has none, and since the metre is co-ordinated with the best system of measures yet in existence while the yard is bound up with the worst, one would think it is the yard which ought to give way. For the metre is practically the forty-millionth part of the circumference of the earth, measured through the poles; that is to say, it is the ten-millionth of the meridian quadrant, or distance from the pole to the equator. This is a most useful aid to memory in making off-hand calculations, and although not absolutely correct (that is out of the question), the error is so slight as to be of no practical importance, being only about a two-hundred-and-eighth part of an inch. This would cause an error of about a hundred yards in the length of Great Britain. I cannot agree with the Commissioners of 1819 when they say: 'There is no practical advantage in having a quantity commensurable to any original quantity existing, or which may be imagined to exist, in nature, except as affording some little encouragement to its common adoption by neighbouring nations.'

But if the yard is in itself nothing in particular, it will surely be found to bear some definite commensurable relation to our unit of weight, the pound avoirdupois. Let us see whether this is the case. We are told by the report of 1819 that 19 cubic inches of distilled water at 50° Fahr. shall be exactly 10 Troy ounces or 4,800 grains, and that 7,000 such grains shall make an avoirdupois pound, and that 10 such pounds of water at 62° Fahr. shall be a gallon. Mark the change of temperature. The grain is first determined by weighing 19 cubic inches of water at 50°. We now have 4,800 grains of water; adding 2,200 more at the same temperature, we have 7,000 grains, and this is a pound avoirdupois. Multiply it by 10, and we have 70,000 grains of water at 50°. And this shall be a gallon. Stay, it does not quite fill the gallon which we have here ready made; very nearly, but not quite. Then let us warm it up till it does fill this old pot, and then note the temperature. We find it is now at 62°, and the pot is quite full. The question arises, what is a grain? Is it a weight or a bulk measure? The actual pot used for this juggle was not a gallon, but the eighth part of the intended gallon. It was an old pint pot which had been lying in the Exchequer time out of mind, and which did, as a fact, hold just 8,750 grains of water if warmed up to 62° Fahr. We have thus wriggled from a measure of water, namely 19 cubic inches. weighing exactly 10 Troy ounces to another measure of water, namely the old pint pot, weighing exactly 20 avoirdupois ounces; and the only shuffle required to show how providentially the pint pot was related to the inch, and thereby to the sacred yard (namely, the length of Henry I.'s arm) was the warming up of the water from 50° to 62° Fahrenheit. In adopting this method for finding the cubic contents of a tank, cistern, or reservoir containing a given weight of water, we must not forget to reverse the process and cool our water down again, or an error will result.

This is what came of the struggle in 1819 to co-ordinate our measures of length, bulk, and weight. According to the metric system, this is effected more easily, and perhaps more ingenuously, by the simple process of calling the cubic metre a kilolitre, and by calling the weight of pure water contained in that measure a millier, the water from first to last being at the temperature of melting ice. If the thing is worth doing at all—certainly, if it is worth the tedious, albeit ingenious, legerdemain of Thomas Young and Captain Kater—it is worth doing well and thoroughly. What is true of our weight measures is equally true of our bulk measures; there is no definite or commensurable relation between the yard and the gallon or bushel; nor, when we come to surface measures, can any one say what is the length of a side of a square acre: it is somewhere between 69 and 70 yards, but no one knows exactly what.

With regard to precision our own standards leave little to be desired; they are probably as accurate as science and the art of instrument-making will admit of. Within the present century precision has made immense strides under all systems, and therefore it cannot be put forward as an argument in favour of any particular system. When we learn that the imperial standard pound is well within a thousandth part of a grain of what it pretends to be, and that the four parliamentary copies are all within two thousandths of a grain of their average; and that the precise length of the imperial standard yard is known within a few ten-millionths of an inch, we cannot well complain of any carelessness on this score. Let us contrast our platinum pound and its four copies with the old pint, quart, gallon, and bushel kept at the Exchequer since the reign of Queen Elizabeth. They were measured by Dr. Wollaston about seventy years ago, and were found, after being translated into gallons, to stand to each other in the following relations :—

Pint .	. × 8 = 276·9	cubic inches.
Quart	. × 4 = 279·3	,,
Gallon	. × 1 = 270·4	,,
Bushel	. × ⅛ = 266·1	,,

showing variations in the Exchequer standards themselves amounting to over 13 cubic inches to the gallon. Compare the following :—

Imperial standard pound weighs 7,000·000520 grains.
Copy No. 1   ,,   7,000·000758   ,,
Copy No. 2   ,,   6,999·999358   ,,
Copy No. 3   ,,   6,999·998458   ,,
Copy No. 4   ,,   6,999·997098   ,,

The largest is less than one two-thousandth part of a grain over the average weight of the five, and the smallest is about one two-thousandth of a grain under.

What the shrinkage may have been in the length of Henry I.'s arm between getting up in the morning and going to bed at night, between winter and summer, and between youth and old age, we are not told, nor whether the yard was measured from the end of his middle finger, when the nail was cut, to the shoulder joint or to the top button of his waistcoat; but of this we may rest assured—the standard yard was a rough kind of measure. On passing to the extant statutes, we find no reference to the king's arm, but in 1324 it was ordained 'that three barleycorns round and dry make an inch, twelve inches a foot, three feet a yard, five and a half yards a perch, and forty perches in length and four in breadth an acre.' Again, it is enacted by 51 Hen. III. 'that an English penny called a sterling, round and without clipping, shall weigh thirty-two wheat corns, in the midst of the ear; and twenty pence do make an ounce, and twelve ounces one pound, and eight pounds do make a gallon of wine, and eight gallons of wine do make a London bushel, which is the eighth part of a quarter.' It is said that in the reign of Henry VII. the new bushel was actually constructed on this method. But with so loose and vague a unit to start with, and having regard to the multiplication of effects, it is permissible to doubt that such was really the case. It is far more probable that the grains were selected by a standard measure than that the measure was taken from any particular grain, however well grown in the midst of the ear. We must set this improvement in precision down to the immense strides made in modern times in the manufacture of scientific instruments of accuracy, and we must not forget that a like precision has been attained in the standards of France, Germany, and several other countries, and consequently cannot be urged in defence of our English system.

But apart from precision, has our English system anything whatever to recommend it? We have seen that its units of length, area, bulk, and weight are unco-ordinated, and bear no definite or commensurable relation to each other. In what, then, do its virtues consist? In the method of its sub-divisions? This contention has not been put forward since 1819, when the Commissioners reported that 'the sub-divisions of weights and measures at present employed in this country appear to be far more convenient for practical purposes than the decimal scale, which might, perhaps, be preferred by some persons for making calculations with quantities already determined. But the power of expressing a third, a fourth, and a sixth of a foot in inches without a fraction is a peculiar advantage in the duodecimal scale, and for the operations of weighing and of measuring capacities the continual division by two renders it practicable to make up any given quantity with the smallest possible number of standard weights or measures, and is far preferable in this respect to any decimal scale. We would, therefore, recommend that all the multiples and sub-divisions of the standard to be adopted should retain the same relative proportions to each other as are at present in general use.' Anything more inconsequential than this conclusion it would be difficult to formulate. Because the duodecimal scale possesses the peculiar advantage of expressing thirds, fourths, and sixths without a fraction, therefore we should retain the same relative proportions of our measures as are at present in general use. Seeing that our system is not based on a duodecimal scale, and that the foot was grafted on to our system by the Romans, and seeing, further, that with the exception of this and of the ounce Troy (another Roman importation, which is now happily well nigh extinct), we have no measures in any of our series divided into twelfths, it is no wonder that this plea has fallen into desuetude of late. Take our land measures; here, $5\frac{1}{2}$ yards make a pole, 40 poles make a furlong, and 8 furlongs make a mile. Surely, if our English system can still be said to rest on any particular scale, it is the binary; 2 pints make a quart, 4 quarts make a gallon, 8 gallons make a bushel, 8 bushels make a quarter, and so on. We have seen that 8 furlongs make a mile; then 8 pounds make a stone in the meat market; 8 larger stones make a hundredweight. Again, 8 fluid drachms make an apothecary's ounce; 16 avoirdupois ounces make a pound. Indeed, when we turn to the chapter dealing with the restoration of the early English system we shall find that it was based almost entirely on the binary scale. It is probable that not one of the distinguished men who signed the report of 1819 had any experience of the immense advantage of a decimal system. Still less does the idea of co-ordinating the national measures seem to have occurred to them as of much importance; they concerned themselves, as we have seen, chiefly with the convenience of retail tradesmen in making up their parcels. In order to see what importance they attach to the co-ordination of measures, we have only to listen to their opinion on the best way of defining a measure of capacity. 'It would in our opinion be advisable to invert the more natural order of proceeding, and to define the measures of capacity rather from the weight of the water they are capable of containing than from their solid contents in space.' They then proceed to declare the accidental fact that nineteen cubic inches of pure water at 50° happen to weigh ten Troy ounces, and upon that interesting discovery to build up an imperial standard pound in the extraordinary mode already described.

Mr. Chisholm's curious defence of our imperial system, with its unscientific variety of scales, is easily met. Such as they are, he says, it should be remembered that they have been adopted because they were found to be 'practically adapted to the wants and habits of the English people.' Nothing of the kind. They were adopted, as we shall see when we come to the chapter dealing with the history of our measures, for very different reasons. To take a few instances, the Roman foot and mile were adopted because Cæsar conquered England, and his successors were under the necessity of making military roads. The furlong was adopted because it was about the length which an ordinary team of oxen could pull a plough at a single burst. The gallon was adopted because the Roman soldiers could not understand the measures used by the Gauls and Britons, and preferred to purchase their beer and shrub by the helmetful. The firkin, kilderkin, and hogshead were adopted because the Dutch merchants happened to be already provided with casks of those capacities. The pole was adopted because it happened to be the length of the pole in ordinary use, and was therefore the handiest measure for rough field surveying.

Since 1854 there have been no two opinions as to the relative merits of the decimal scale and our own or any other. The commissioners appointed in 1843 pointed out, in their Report of 1854, this change in public opinion. 'Referring more particularly to chapters v. and vi. of the Report of 1841, which recommend a decimal system of coinage and a sanction of decimal systems of weights and measures to a certain extent, we propose with the utmost confidence to carry out these recommendations, and even to advance further, adopting some proposals which have originated with other bodies. We wish to state our opinion that, in reference to the decimal scale generally, the public mind is very greatly changed, and that the introduction of the decimal system will now be very easy in respect to many points which a few years ago would have offered great difficulty.' They proceed to express a hope 'that there is now a prospect of attaining the long-desired simplification of the British system of weights, and for this reason (in addition to those founded on the extreme convenience of a decimal scale in any special system of weights) we are anxious that every facility should be given to the introduction of the decimal scale based on the pound weight.'

The Report was signed by the Astronomer Royal, Sir John Herschel, Professor Miller, Mr. F. Baily, Sir John Lubbock, and half a dozen others, and the name of the Earl of Rosse was subsequently added; and from that day to this no scientific man of any eminence has denied that the introduction of a decimal system of some kind or other is merely a question of time. How long we shall continue to wallow among perches and roods, quarts and reputed quarts, hogsheads and kilderkins, quarters that are quarters of nothing in particular, pounds Troy and pounds avoirdupois, drams and fluid drachms, stones and butcher's stones, hundredweights that are not a hundred anything, tons and tuns, miles and knots, pounds sterling weighing $123·27447$ grains of standard gold, eleven twelfths fine, and all the rest of the chaotic jumble, depends solely upon the time which it takes to present the public with a complete and satisfactory metric system.

CHAPTER IV

MODERN FRENCH SYSTEM

LET us turn for a moment to the French system legally established in most of the civilised countries of Europe. The unit of length is, as we have seen, the metre, which is the ten-millionth part of the meridian quadrant, and which measures about 39·37 inches English. The unit of surface measure is the are, which is 10 metres square, and therefore contains 100 square metres; the unit of bulk measure is the litre, which is a cube whose base is one-tenth of a metre; the unit of weight measure is the gramme, or the weight of a cube of pure water whose base is the hundredth part of a metre. Having got their unit in each of the series, the French adopt the Latin names for ten, a hundred, and a thousand, and use them as prefixes to signify the tenth part, the hundredth part, and the thousandth part of the unit. Thus we have the metre, the decimetre or tenth of a metre, the centimetre or hundredth, and the millimetre or thousandth, part of a metre; similarly, we have the decilitre, the centilitre, and the millilitre. Also, the decigramme, the centigramme, and the milligramme.

Then they take the Greek words for ten, a hundred, a thousand, and ten thousand, and use these as prefixes to denote the multiples of the unit. Thus we have the decametre or ten metre, the hectometre or hundred metre, the kilometre or thousand metre, and the myriametre or ten thousand metre. Similarly, we have the decalitre, &c.; also, the decagramme, the hectogramme, and so forth.

At first sight.this arrangement appears as simple and as scientific as could be desired; but let us put the four series in four columns, with the units on the same line, the deci's on the line above, and the deca's on the line below, and so on, all in order. Here they are :—

Millimetre		Millilitre	Milligramme
Centimetre	Centiare	Centilitre	Centigramme
Decimetre		Decilitre	Decigramme
Metre	Are	Litre	Gramme
Decametre		Decalitre	Decagramme
Hectometre	Hectare	Hectolitre	Hectogramme
Kilometre		Kilolitre	Kilogramme
Myriametre	Myriare		Myriagramme
			Quintal
			Millier

In this arrangement of the table it will be seen that the four units of length, area, bulk, and weight are set in a line, with their sub-multiples and multiples arranged in the order of nomenclature; that is to say, the tenths are all in one row, the hundredths in another, and the thousandths in a third; those in the same line all having the same Latin prefix, while the multiples are similarly arranged below the unit line with their Greek prefixes.

But the first thing to attract adverse notice is the fact that the four units do not bear the relation each to each that we should expect. The are is not the square of the metre; the litre is not the cube of the metre nor of the are; and the gramme is not the weight of a litre of water, nor of a cubic metre of water. When we come to the area series we do not find any milliare, deciare, decare, or kiliare; presumably because such superficial measures would be of no conceivable utility, having no measurable base. This objection, however, is not held to militate against the introduction of fractional bulk measures and weight measures, which share this defect. For example, we have the hectolitre, a familiar wine measure, which is a cube whose base, expressed in decimetres, is the cube root of 100, whatever that may be.

Turning to the first line, we find the milligramme, which is the weight of a cubic millimetre of water, but not the weight of a millilitre. A millilitre of water weighs a gramme, and the length of its base is a centimetre. Taking the weight series, when we reach the myriagramme, a weight of about 22 English pounds, we have exhausted our Greek prefixes, and an auxiliary nomenclature must be resorted to. And a weight equal to about an English ton is called a millier, a term which ought to signify, according to the French system, the thousandth part of an -er.

But although the co-ordination of the four series is by no means obvious when arranged as above, it is possible to disregard the Latin and Greek prefixes, and to arrange the table in such a way as to bring the true relations of the four series into light. The following table is so arranged:

Millimetre			Milligramme
Centimetre		Millilitre	Gramme
Decimetre		Litre	Kilogramme
Metre	Centiare	Kilolitre	Millier
Decametre	Are		
Hectometre	Hectare		
Kilometre	Myriare		
Myriametre			

But if we adopt this arrangement, the special virtue of the French nomenclature disappears. Moreover, the whole system is meagre in the extreme. We have only 3 bulk measures, 4 area measures, and 4 weight measures, of which one fails to comply with the conditions of the nomenclature. Nevertheless, this arrangement is decidedly preferable to the former one. One can now see at a glance the relation between any two of the measures on the same line. For example, the square of the metre is the centiare, the cube is the kilolitre, and the weight of a kilolitre of water is a millier. The cube of the centimetre is the millilitre, and the weight of a millilitre of water is a gramme. And the weight of a litre of water (that is, a cubic decimetre) is a kilogramme. But even now there is only one base which runs throughout all four series—viz. that of the metre, the square of which is the centiare in the area series, the cube of which is the kilolitre in the bulk series, and the weight of which cube in water is the millier.

So far from being of any assistance to the memory in this form of the tables, the Latin and Greek prefixes are a prolific source of confusion. The much belauded nomenclature becomes not only useless but mischievous. The area measures do not extend to surfaces of sufficient smallness, and the weight measures do not extend to weights of sufficient greatness. The largest of the orthodox weights is about 1 stone 8 pounds English.

Again, consider the unwieldy length of the names. They actually average between eight and nine letters. A dozen of them contain ten or more letters each. If these could be easily abbreviated it would matter less. But how are we to abbreviate milligramme and millier so as to distinguish between them? And remember they are both in the weight series. There are eight names beginning with 'm,' four beginning with 'milli,' three with 'deci,' and three with 'deca.' The Royal Commission of 1869 said there was little fear of confusing the decigramme with the decagramme, but they contrived to do it twice themselves in the very tables they furnished as an illustration.

It has been contended by some that, apart from the scientific beauty of the French nomenclature, with its Latin and Greek prefixes, it is well that all nations should use the same names. Now, there seems to be no more reason for using the same measure names than there does for using the same quality names. A universal language would doubtless be convenient, but until we get it there can be little good in forcing English people to talk French, or French people English. If the measures are the same, the names of the measures can be learnt in an hour or less by any person of average intelligence. Those who trade with foreigners are never long in learning the foreign names of the articles in which they deal. Surely this argument is about the feeblest that was ever put forward as a reason for asking a nation like the English to adopt a nomenclature which is admittedly bad.

There is no need to lay stress on the meaninglessness of the French names, because this is really not of very much importance. Their denotation is soon learnt by experience. But it may be well to point out that, inasmuch as children usually learn their tables of measures long before they learn either Latin or Greek, there is nothing gained, so far as education is concerned, by the use of the classical prefixes. After all, it is as easy to learn that such a word as yasp, for instance, means a thousand grammes, as it is to learn that kilo denotes a thousand anythings. Indeed, it may be doubted whether even a scholar would perceive at a glance that so barbarous a word as kilogramme was intended to mean chiliogramme. Let us karitably admit that he might.

The philosophers, with their usual disregard of human nature, considered a single unit of measure in each series to be sufficient, not only in fact but in name. Now, it is clear that persons comparing the size of type, the thickness of wires, or the length of insects' wings, must, if they think in terms of the metre, use figures consisting of a row of decimals. No one can form any clear idea of one-thousandth of a yard or of a metre. On the other hand, persons wishing to describe the quantity of earth to be removed in the construction of a railway tunnel find the figures too large to be comprehended. Tens of millions of cubic yards is an expression which conveys no definite conception to the ordinary mind. The proper course is to take the most convenient unit for the purpose, and treat it as an integer.

CHAPTER V

A SURVEY OF ENGLISH MEASURES

I NOW propose to meet the arguments of those defenders of our present system of measures who contend that they are suited to our multifarious wants and requirements because they have spontaneously grown up in our midst. In order to do this let us pass in review most of our leading customary measures of length, area, bulk, weight, and value, and find out how they really did originate. We will begin with the length measures.

The names of our principal measures of length are the inch, foot, yard, pole, furlong, and mile. In addition to these we have the barley-corn, hand, span, ell, fathom, link, and chain. Inasmuch as our area measures can hardly be said to have a separate nomenclature, it may be as well to consider them at the same time. We have the square inch, the square foot, the square yard, and the square pole, without any distinctive names of their own, and in addition to these we have the rood and the acre.

Of these the inch, foot, and mile were brought over by the Romans, and forced upon the British in spite of the fact that they bore no commensurable relation to the then existing measures of length. The yard, the furlong, the rood, and the acre may be said to be of home growth, while the link and chain are a comparatively recent invention of land surveyors. The relations subsisting between the several members of this jumble are about as complicated and objectless as could have been devised by any of Mr. Gilbert's most eccentric statesmen. How the Commissioners who signed the Report of 1869 brought themselves to believe that there must be something supremely convenient in a pole of $5\frac{1}{2}$ yards passes the wit of man to conceive. It is also difficult to understand why a mile of ground should be 265 yards less than a mile of water. It would puzzle some people to explain why a quarter of an acre should have a name at all, and those who applaud our binary system of subdivision might be asked how it came about that the yard was divided into three feet. But we will begin our historical survey with the famous unit itself, our wonderful yard.

Etymologists may be able to tell us how a yard came to be the precise length which it now is. Was it the length of a man's ordinary stride? or was it the length of his staff? or was the one derived from the other, and, if so, which came first? Of the nun who could not bear to see her dog thrashed, Chaucer says:

> But sore wept she if on of hem were dede,
> Or if men smote it with a yerde smerte.

And the same writer uses the word thus:

> Me mette, howe that I romed up and down
> Within our yerde, wher as I saw a beste,
> Was like an hound, and wold han made arreste
> Upon my body, and han had me ded.

It matters little to our present inquiry whether the garden or yard was a guarded plot from which the stakes with which it was guarded took their name, or whether the stakes or yards gave their name to the plot so yarded round. It may be noted that the word 'gird' means both to strike or gird at and to encircle with a girth or girdle. Similarly, a ward is an enclosed place and we ward off a blow. The point for us is that a mete-yard was a measuring rod long before the time of Chaucer. Indeed, we have the Anglo-Saxon mete-geard. Before the introduction of the metric system into Germany, their unit of length was called a stab, which is our word staff. It varied in different localities from 1·16975 English yards at Freiborg to 1·31236 yards at Frankfort. In short, it varied as walking sticks do between about a metre and a yard. Possibly the original yard was also a very indefinite measure of about the length of a walking staff. We shall see.

Using the binary system of division as the Anglo-Saxons did, we should expect to find the yard, the half-yard, the quarter-yard, the eighth of a yard, and even the sixteenth of a yard. So we do, and the last of these measures is called a nail. What kind of nail it was which gave its name to this precise measure I cannot tell. It is about the length of the first two joints of the index finger. But that would never be called the nail. It was probably some nail, needle, or nagel in common use. The idea of sub-dividing the yard into thirds would never have occurred to our ancestors. But, finding the Roman foot in common use, and comparing it with their yard, they ascertained by measurement that a yard contained about three feet. The Romans, as we know, divided all their chief measures into twelfths called *unciæ*. The twelfth part of a foot or *pes* was called an *uncia*, which is our word inch. The twelfth part of the Roman pound was also called *uncia*, which is our word ounce. The twelfth part of an hour, or a period of five minutes, was also called an *uncia*. Again, the twelfth part of the Roman acre, or *jugerum*, was also called an *uncia*. Possibly the familiar expression 'a hunch of bread' has the same derivation. It is surely rather absurd to tell us that, because the ancient Romans found it convenient to measure distances by the simple process of putting one foot before the other, and because they had got into the habit of dividing all their measures into twelfths, there must be something about the foot and inch peculiarly adapted to the wants and purposes of Englishmen. There is much more to be said in favour of the yard, for when once the walking staff had been selected as a convenient rough unit of measure, it was easy to construct it of a precise length; and then for small measures nothing was easier than to take a piece of string of that exact length and to fold it in two for the half yard, re-fold it again for the quarter, and so on.

It may be worth notice, *en passant*, that all the Greek body measures are a shade larger than our own, and that all the Roman equivalents are a shade smaller. The Roman *palmus* was just under three English inches, and the Greek παλαιστη was just over three inches; the Roman *cubitus* is just under, and the Greek πηχυς is just over, the English cubit or half yard. Our fathom is as much rope as an average sailor can tell out across his chest with both arms outstretched. The word is allied to the German *faden*, a rope or line. Here again we are just beaten by the Greeks, whose ὀργυια was four-fifths of an inch over six English feet or a fathom.

The rest of our national length measures—namely, the rod, pole, or perch, the link, chain, furlong, and mile—all hinge upon the furlong, and they are best considered in connection with our land or surface-measures. Otherwise there is no explaining on rational principles the adoption of such a measure as 5¼ yards for a unit. Nor can we in any other way explain how the mile, which is the Roman *millepassuum*, came to be divided into seventeen hundred and sixty parts instead of a thousand. For it must not be supposed that our mile has grown quite so much as these figures would lead us to suppose since the Romans brought it over. On the contrary, it is only about 142 yards longer than it was when they left. That is to say, the old *millepassuum* is only about a twelfth less than our modern mile. No, the explanation is that the old yard was more tenacious of life in the fields than any of the Roman divisions of the mile.

Now let us look at our surface measures. First of all, I must protest against the use of the expression 'square measures.' Our ancestors did not trouble themselves about square measures. All they wanted was a system of convenient land measures. Take the acre; how would you describe it? A square whose side is——? Well, it is less than 70 yards and it is more than 69 yards. In short, it is a most unconscionable fraction. The rood, again, if regarded as a square, has a side of no determinate length; it is somewhere between 35 yards and 34¼. Then comes the square perch; it is difficult to understand why our plough poles should have been 16 feet 6 inches long, while the Roman plough pole was only 10 feet long. Was it that our uncultivated and rough fields required an extra pair of oxen to drag the plough through them? If so, this will also account for the fact that the Roman furrow length was less than half our furrow-long, or furlong. That is to say, our customary team could do a furrow of 220 yards at a single spurt, whereas the Roman team could only manage 80. Mr. Seebohm, in his *English Village Community*, says: 'The presence of the team of 8 oxen in Wales and Scotland as well as in England, and the mention of teams of 6 and 8 oxen in the Vedas, as used by Aryan husbandmen in the East, centuries later, makes it possible, if not probable, that the Romans in this instance, as in so many others, adopted and adapted to their purpose a practice which they found already at work, connected perhaps with a heavier soil and a clumsier plough than they were used to south of the Alps.' Commenting upon this passage, Mr. Gomme, in his *Village Community*, observes: 'It does not seem clear that the conclusion here arrived at is sufficient for the facts. The question as to whether the Romans adopted the eight-oxen plough team into their system cannot depend upon the wide prevalence of such a team in Aryan countries. The conclusion from such evidence is rather that the plough team as an essential in primitive agricultures lies altogether outside the sphere of Roman influences one way or the other.'

Now this burst or spurt is the natural unit of land measure, because it would obviously be bad economy to plan out the fields in lengths which would necessitate the turning of the plough before the burst was exhausted. There is nothing mystic about the number 220. It is merely the number found by direct measurement to express the ratio between the yard and the natural furrow-long. Or, more likely, the furrow was first measured by the plough pole, 40 of which were found to go to the average furlong, and then the plough pole was found to measure about $5\frac{1}{2}$ yards. The precise length of a furrow is not a matter of a yard or two. Sometimes the guessed furrow would measure 39 poles odd, sometimes it would measure 40 odd, and sometimes a trifle more or less than either. In the end, for the purposes of precise measurement, the ratio would settle down to exactly 40 poles to the furlong. Similarly, poles would vary in length, though not much. But eventually the legal pole measure would settle down to the nearest workable multiple of a yard—namely, $5\frac{1}{2}$.

The Romans were more under the influence of mathematical theories, and severe was the struggle in Italy between the decimal and duodecimal systems. Having got their ten-foot pole, they would try to fit the natural furrow to some duodecimal multiple of that pole. They would have wished to make it 12 poles long, but that was far too uneconomical; so they had to make it 24. But the width of their acre or *jugerum* was of far less importance, so they carefully made it twelve poles wide. Consequently, the Roman *jugerum* was 240 feet by 120 feet, or 28,800 square feet. This is about five-eighths of an English acre, and it tends to show how artificial the measure was when we find that two *jugera* were reckoned to each citizen as heritable property. Thus the *heredium* was about an English acre and a quarter. The Roman name for the pole was *pertica*, which is our word perch. *Jugum* means a yoke, and Varro says that it meant the quantity of land that a yoke of oxen could plough in a day. Our word acre is simply the Latin *ager*, a field; and whilst its length was dependent on the natural furlong, its width would no doubt be reckoned much in the same way as the *jugerum*, so as to give a total quantity of land such as could be ploughed in a day. Apparently, this was found to be what we call an acre, or a furrow-long by four poles wide. This measure of four poles afterwards came to be called a chain. We must remember that originally the plough had to come back after each furrow had been ploughed, as the reversible ploughshare had not then been invented. Moreover, carts were not made as they are now. They were heavy lumbering things, and when laden it would be an economy of force, especially when land was cheap, to leave a roadway of flat grass land every now and then for the carts to go up and down to collect the produce and to deliver the manure. Thus the acre came to consist of three-quarters of an acre plough land and one quarter grass land. The name given to this quarter strip of grass land would be the obvious one, road or rood. It would be one furrow long and one pole wide, which is just about the width required for a cartway where land is plentiful. This, then, is the rood of which we now speak as though it were a square plot. On the contrary, it is a road 220 yards long by 5½ yards wide—a long narrow strip of grass land. This may help us to explain a passage in Richard Grafton's 'Chronicles': 'And five yards, halfe a perch, or poll, and XL pol in length and thre in bredth an acre of land.' Surely this is incorrectly punctuated, and should read, 'And five yards half, a perch or poll; and XL pol in length and thre in bredth an acre of land.' Meaning that 5½ yards make a perch or poll, and that 40 poles in length and three in bredth make an acre of land. But even so, how does it happen that he makes an acre only 3 instead of 4 poles in breadth? My answer is that he was speaking of the land actually ploughed in an acre, and that he therefore excluded the rood of grass land.

It is clear that these grass roods would traverse each other all over the open fields like the bars in a window-frame, or the lines which separate the squares of a chess-board. They would form crosses at each point where four strips met. Is it not probable that when Christianity came to be preached to our Teutonic ancestors the cross would be spoken of and described by reference to these intersecting roods? Hence, perhaps, the Holy Rood. Our early writers do not seem to use the word 'road' at all in the sense of a way. They always use the word roadway. And when they use the word 'road' it is in the sense of raid. 'Oftentimes also they would make rodes in the night and assault the castles of our camp,' says Goldinge, as late as the sixteenth century.

The Roman *centuria* bore no definite relation to the Roman mile. They did not deal in square miles. As a chance fact, the *centuria* happened to be very nearly a quarter of a square Roman mile, but no pains were taken to adjust them. In England, on the contrary, the mile is exactly 8 furlongs. This could not have come about by accident. Either the old furlong must have been altered to suit the mile, or the old mile must have been altered to suit the furlong. The

Scrupulum or square scrap 100 square feet

10 feet long or pertica or decempeda

Actus   12 perticæ   120 feet long

Half Jugerum or Actus Quadratus
14,400 square feet or 144 scrupula

1½th of Jugerum

2,404 square feet

J U G E R U M

24 poles by 12 poles,
288 scrupula, or 28,800 square feet

24 scrupula

Uncia

Actus Quadratus

question is, which? Now, our present mile, though retaining its old signification of a thousand somethings, is not in fact a thousand anythings. But when it came over to us it was a thousand Roman paces. Therefore, it must have begun business here as 5,000 Roman feet. It must have grown no less than 426 feet, or 142 yards. The only plausible explanation of this is that it has been drawn out so as to fit in with the English binary system, and so to do duty as 8 furlongs. Just as 8 pints make a gallon, and 8 gallons make a bushel, so must 8 furlongs make a mile. Now the furlong could not be altered. The people will not allow any tampering with old land measures. Therefore the mile must be drawn out from 5,000 Roman feet to 5,439. The square mile has never been popular with us any more than with the Romans, and we still speak of the acreage of a country, and even express it in acres.

Returning to the rood, the crude notion that this area-measure took its name from the fact that, looked at from a distance, these narrow strips of grass-land bore some resemblance to rods is disposed of at once by two considerations. One is the obvious connection between the words 'road' and 'ride': the road is what is 'rode' on, and making a raid is merely cutting a road through hostile country. The other consideration is that the Roman *actus* got its name in precisely the same way. *Actus* is literally a drive or cartway such as beasts can be driven along. As Pliny says, 'Actus vocabatur in quo boves agerentur.' If he had stopped there his definition would have done him credit; but he goes on to add, 'cum aratro, uno impetu justo.' Surely this is a mistake, or else the Roman furrow, instead of being the whole length of the *jugerum*, would have been only half that length —an absurdly short furrow, especially in the olden time. It may be that the word *actus* was first used to denote the roadway at the head of the *jugerum*, and then extended so as to denote also the grass roods between the *jugera* lengthwise.

On referring to the plan of the *jugerum*, it will be seen at a glance that it is twice as long as it is broad; that it consists of twelve long strips, called *unciæ*, each of which can be divided into twenty-four little squares, called *scrupula* or scraps of land, 10 feet by 10; and that the *jugerum* can be cut into two squares measuring an *actus* each way. This *half-jugerum* was called an *actus-quadratus*. Each square *actus* would thus contain 144 *scrupula*. It would resemble a chess-board with twelve squares each way. This is a good instance of the duodecimal system prevailing in Rome. If you lay two *jugera* side by side, you get another square called an *heredium*. One hundred of these large squares formed another chess-board of ten squares each way, called a *centuria*. And this is a good example of the triumph of the decimal system. Four of these big chess-boards form a still larger square, called *saltus*; but this area was of no practical value, and was only a feeble attempt to get an approximation to the squared *millepassuum*; and a very feeble attempt it was, for the mile contained 25,000,000 square feet, while the *saltus* contained a trifle over 23,000,000, or, exactly, 23,040,000 square feet.

Turning now to the plan of the English square mile, it will be seen that it forms a chess-board of sixty-four squares, the regulation number of a true chess-board. So that anyone with such a thing in his possession can easily furnish himself with a plan of an English square mile and its sub-divisions. Each square is a square furlong. This much is already done on the board. Let him now divide each square into ten strips, and each such narrow strip is an acre. By dividing one of these acre strips into four still narrower strips, and colouring one of the four green, he will have the rood. And by cutting up the rood into forty little equal segments, each of which he will find to be a square, he will see before him forty square perches. If his board happens to measure 32 quils or centimetres each way—a very ordinary size—then each square perch will be represented by an en or square millimetre; the rood, the acre, and the square furlong will each of them be 4 quils long; the rood will be 1 jot or millimetre wide, the acre 4, and the square furlong 40. The square perches will be quite visible to the unaided eye, and indeed countable, and the whole board will give a better idea of the relative magnitudes of our English land measures than any number of tables.

It will, I think, be admitted that our old system of land-measures is not so contemptible as is sometimes supposed. It is certainly a pity that there happened to be $5\frac{1}{2}$ yards in a pole. If there had been 4 or 8, it would have been a lucky accident. But neither the yard nor the furlong could give way, and so it came about that the $5\frac{1}{2}$ yard pole was found to be the best to work in with both these measures. The evil consequence of this is more manifest when we come to the square pole, which, of course, measures the uncomfortable number, $30\frac{1}{4}$ square yards. It would tax the resources of a surveyor to arrange 30 mats, each of a square yard, and a quarter of a mat, in the form of a square perch. He would have to put 25 mats in the form of a square, cut 5 of the mats in two, and place the 10 half mats along two of the sides of his square, and then stick the quarter mat up in the empty corner. This is a dismal process; and it certainly does not lend itself to rapidity of calculation. It is only fair to our forefathers to recollect that the idea of a square rood or of a square acre no more entered into their minds than the idea of a square egg.

It may be permissible at this point to suggest by way of contribution to the perennial discussion on the history of chess, that in all probability its present form at any rate was settled by the northern nations, though it seems to have come in a crude form from the East. The Romans would certainly have constructed a board with 144 squares in their early days, and with 100 squares in later times; but the northerners, with their instinctive preference for the binary system, would just as certainly have chosen either 16 or 64 or 256 squares, and in all probability they did choose 64. It may also be mentioned here that our cricket pitch is exactly the length of an acre's width, namely, 22 yards or a chain, just as it was when the game was played after Divine service in God's Acre. I am inclined to suspect that the original distance between the goals in football was originally 1 furlong, or the length of God's Acre. It will be remembered that John Knox used always to take part in this game after church on Sundays.

When the old pole was more or less supplanted as a measuring instrument by a chain, an attempt at a decimal system seems to have been made. Indeed, the shape of the acre suggests it. Finding the acre already ten times as long as it was broad, the width was naturally taken as the unit, and called simply a chain, and the length was, of course, ten chains. The chain was then divided into links, and the question was how many links there should be in the chain. It is well that there should be the same number of links in every chain, so that one has only to count the links in order to know whether or not it is of the full chain length. The wonder is that 22 was not the number chosen. However, it was decided to have a hundred links in the full chain of an acre width. Perhaps links a yard long would make too unwieldy a chain to drag about, but surely a chain with links about $3\frac{1}{2}$ inches long would be handy enough. The present link being 7·92 inches is, taken by itself, of no use whatever as a measure.

We are now in possession of the complete series:—

12 lines make a inch,
7·92 inches make a link,
12 inches make a foot,
3 feet make a yard,
5·5 yards make a pole,
40 poles make a furlong,
8 furlongs make a mile.

Consequently our table of surface measures presents the following wild appearance:—

144 square inches make 1 square foot,
9 square feet make 1 square yard,
30·25 square yards make 1 square perch,
40 square perches make 1 rood,
4 roods make 1 acre,
10 acres make 1 square furlong,
64 square furlongs make 1 square mile.

SQUARE MILE, SQUARE FURLONG, ACRE, ROOD, AND PERCH
(Scale, 2 quils to the furlong; 2 jots to the chain.)

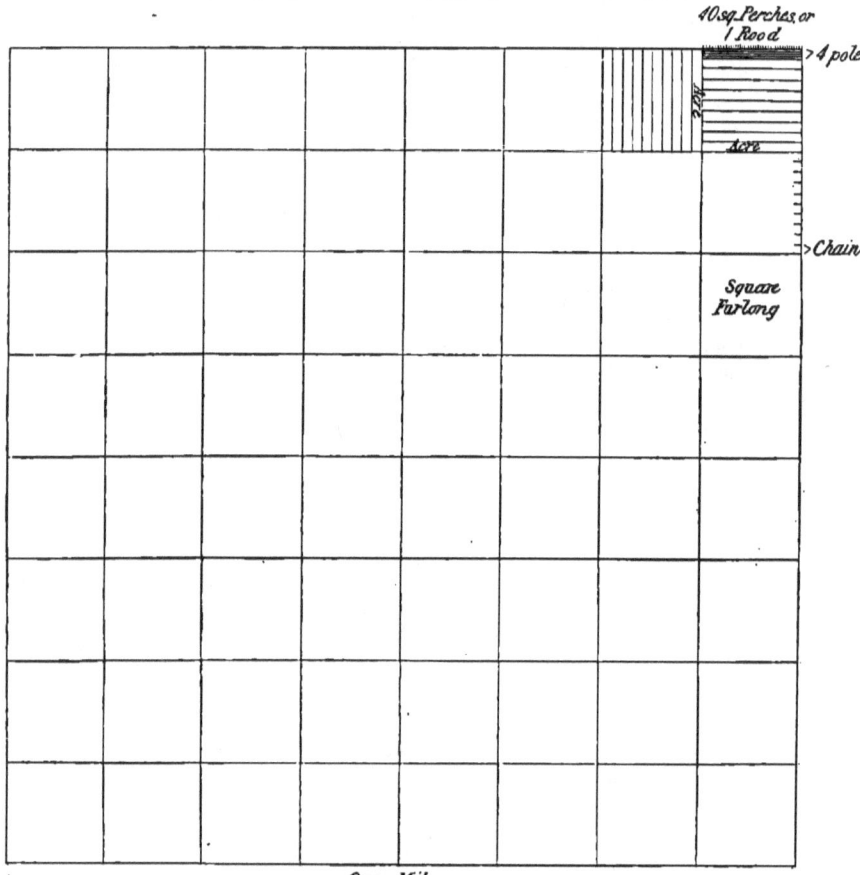

See Plate VIII. for Acre on larger Scale

I think, having regard to the history of these measures of length and surface, very few will be found bold enough to defend them on the ground that they have grown up spontaneously to meet the requirements of the English people. The old and forgotten plough pole, the thousand Roman paces, the long grass road through the open fields, the furrow suited to the ox team—all these have about as much connection with the present needs of Englishmen as the prehensile tail of an ape.

But perhaps we shall find more to justify this optimistic view when we come to consider the measures of bulk. First, there is the distinction between dry and liquid measures. Secondly, there is the hopeless endeavour to co-ordinate some half-dozen different systems. We have Roman measures with Roman names, Norman measures with French names, Anglo-Saxon measures with English names, and, above all, Dutch measures with names that have been mangled out of all recognition.

We have tuns and butts, vats and quarters, hogsheads, coombs, barrels, sacks, pipes, strikes, kilderkins, bushels, ankers, firkins, pecks, gallons, quarts, pottles, pints, nipkins, gills, and drachms. Several of our bulk and weight measures are British, and presumably date back to pre-Roman times. Such are the tun, the tod, and the coomb. Again, the following came over from Holland: the last, the hogshead, the kilderkin, the firkin, the stoup. We are indebted to the Normans for the quarter, the bushel, and the quart. The Romans bequeathed to us the gallon, and, perhaps, the gill. Of pure English names there are but few: the pipe, the vat, the sack, the peck, and the pint. How all these measures are related to one another is a question which we have all solved in our infancy, and afterwards forgotten. I propose to inquire into the historical origin of some of them, as I did into that of our measures of length and surface.

It is difficult to know where to begin. I have a reason for starting with the pipe. Pipe is a Saxon word, meaning a cylinder. The English pipe was declared to contain 126 gallons by an Act of Henry VI.; but, as I shall show, at a still earlier date it must have contained

128 gallons. Now the pipe office, or office of the Clerk of the Pipe, was an ancient department of the Court of Exchequer. The meaning of its name was already forgotten in the time of Francis Bacon, who gives us rather a whimsical explanation of it. Says he, 'these be at last brought into that office of Her Majesty's Exchequer which we by metaphor do call the pipe, as the civilians do by a like translation name it *fiscus*, a casket or bag; because the whole receipt is finally conveyed into it by means of divers small pipes or quills, as it were water into a great head or cistern.' The notion of a State office obtaining its name by a far-fetched metaphor is quaint. Surely, the Clerk of the Pipe was none other than what was afterwards called the Warden of the Standards; except that his functions were far more considerable and important, for he had to superintend all the surface measuring and surveying of crown lands, and to make out the leases of the same, and the accounts of sheriffs, besides much else where the standard measures were required. There is reason to believe, as I shall show, that the pipe was the early English unit of bulk measure; and the ancient office would be very naturally styled the pipe office, after the name of the standard. It was abolished in the reign of William IV., and its records were transferred to the custody of the King's Remembrancer of the Exchequer. It would seem as though foreign merchants, being constrained to use our pipes, actually imported from England the requisite hoops and pipe staves for the purpose. Spelman, the antiquary, thinks that the pipe office is so called because the papers were kept in a large pipe or cask. These ancient opinions are valuable only as showing that the history of the pipe was already lost in their day.

I shall show that the old pipe was a cylinder whose height was equal to its diameter. The eighth part of a pipe of wheat was a boll, which for purposes of exact measurement was also a cylinder of half the height and half the diameter of the pipe. The eighth part of this again was called a peck, which was a cylinder a quarter the height of the pipe and a quarter its diameter. The peck contained eight cans and the can contained eight gills. What the precise height and diameter of the pipe and its sub-multiples were we shall see when we come to the restoration of the early English system. Meantime it should be observed that these five measures so simply related all bear Anglo-Saxon names, unless the gill is the Low Latin *gilla*. The pint or pynt is roughly about as much water or ale as a thirsty worker can drink off at a drain; the word was pronounced so as to rhyme with 'flint,' and it was not till long after the Norman conquest that it obtained its present Frenchified pronunciation. It is derived from *pyndan* meaning to contain or hold or pen-in. The weight of a pynt or pynd of standard wheat bore the same name as the measure, namely pynd; and I shall refer to it more particularly when we come to the consideration of our measures of weight. As a measure of capacity it was half a can; and the measure itself was a cylinder of the same diameter as the can, but one half the height.

Eight pints made a gallon. There is no doubt that the terms quart, gallon, bushel, and quarter came over with the Normans. It is not probable that originally they were precisely equal to the English measures; but they seem to have been forced into the system very much as the Roman mile was altered to suit the furlong. There has been much dispute about the derivation of the word ' gallon '; but I have little doubt that it originally meant a helmet-full. When Western Europe was overrun by Roman soldiers, and when local measures were unknown to the invaders, there would be no handier measure for ale or shrub than the soldier's helmet, for which the Latin is *galea*. The *galeola* is a little helmet-shaped cup; and the word *gallona* also occurs as a measure. Moreover, we find the word in some form or other in all the Latin tongues.

The quart is of course the quarter-gallon. But what was the measure called a quarter? What was it a quarter of? If we refer to our tables we shall find that 5 quarters make a wey or load. But what do 4 quarters make? I think there is little doubt that the name was made to do duty for the quarter of a tun, which was a double pipe—that is to say, a cylinder of the same diameter as the pipe but twice the height. Hence the quarter would be a cylinder of the same diameter as the pipe but half the height. And it probably superseded the fate or vat (German *fass*) as a dry measure. Seeing that 8 gallons make a bushel and 8 bushels make a quarter, it is clear that a quarter was at one time the same measure as a hogshead, but the latter term was used for liquid measure and the former for dry measure, and the two eventually became differentiated.

Let us turn to our liquid measures. Eight pints make a gallon; and at one time eight gallons seem to have made a firkin, which is a Dutch word signifying a fourth part, and it was used to denote the fourth part of a barrel. The half-barrel was called a kinderkin, and meant a baby barrel. It afterwards came to be spelt kilderkin. The double-barrel was called a hogshead. The origin of this name I cannot find; but seeing that it came from the same country, Holland, and that in that country the word *oxhoofd* means an ox-head, it is probable that this is the original word. I am disposed to think that what was meant was not the head or skull of the ox, but that which stood at the head of the ox—namely, his trough. A trough about 4 feet long, 2 wide and 1 deep would contain some such measure of water as that which we now call a hogshead.

The precise relations between these Dutch measures and the English gallon seem to have fallen into oblivion at a very early date. At all events, they were unknown to the Legislature of Henry VI.; for in the Parliament which sat at Westminster in the year 1423 an Act was passed dealing with liquid measures which runs thus:—

'Whereas in old time it was ordained and lawfully used that tuns, pipes, tertians, and hogsheads of wine, barrels of herring and of eels, and butts of salmon coming by way of merchandise into this land out of strange countries and also made in the same land, should be of a certain measure, that is to say, the tun of wine 252 gallons, the pipe 126 gallons, the tertian 84 gallons, the hogshead 63 gallons, the barrel of herring and of eels 30 gallons fully packed, the butt of salmon 84 gallons fully packed; nevertheless, by device and subtlety now late such vessels have been of much less measure, to the great deceit and loss of the King and of his people. . .'

Whereupon the Act proceeds to penalise all such goods which are of short measure according to the ancient scale, and to include the kinderkin and firkin in the list. The penalty was forfeiture to the king, who allowed one-fourth to the informer.

These figures are sufficiently near the multiples of eight to show us clearly what they must have been at a date still earlier than that referred to by Henry as the 'old time.' Let us put Henry's figures and the true binary figures in two parallel columns, and we shall form some idea of the change which had already taken place in his day. The explanation is extremely curious.

Measure.	Original.	Henry VI.
Firkin	8	7
Kilderkin	16	15
Barrel	32	30
Hogshead	64	63
Pipe or Butt	128	126
Tun	256	252

It seems to have occurred to the Clerk of the Pipe, or to some other distinguished mathematical luminary, that the number of gallons contained in the several casks was computed, or should be computed, in scores and units. Thus he says the hogshead should contain three score and three gallons, the tertian four score and four, the pipe six score and six, and the tun twelve score and twelve. This method of reckoning tallies, as will be seen, with the figures in Henry's list; and the method is actually described in an ancient statute.

It is needless to observe that the figures in a third or modern column might be filled up almost anyhow. For instance, the pipe of sherry contains 108 gallons, the pipe of port 115 gallons, the pipe of Madeira 92 gallons, and so forth. The hogshead of claret or Madeira is 46 gallons, of port 57, and of sherry 54. But none of these measures are now recognised by the law, and the goods must be bought and sold by the imperial gallon. Long before the Act of 1824 became law, our liquid measures had lost all meaning. The very gallons themselves varied according to the stuff that was sold. Thus all over England in the provinces a hogshead of wine contained 63 wine gallons; a hogshead of ale or beer contained 51 ale gallons; while in London a hogshead of beer contained 54 beer gallons, and a hogshead of ale contained 48 ale gallons. It will be wearisome to dwell on this dismal wilderness of measures further than is necessary to show into what a tangle 'the subtlety of merchants' had brought our system; and how groundless is the appeal of its advocates to its spontaneous development and peculiar fitness to our needs as a justification of its existence. As a rule, the tendency was for the measures to grow smaller, as we see from Henry's wail; but in some cases the reverse process was set up. For example, it appears from an Act passed in the reign of Henry V. that the cauldron of coals continued to grow bigger and bigger, to the great grief and sorrow of the king. The preamble to this Act runs thus:—

'Whereas of every cauldron of sea coals which he or shall be sold to people not franchised in the port of the town of Newcastle-upon-Tyne two pence be due to the King's Customs, and in the same port be certain vessels called keels, by which such coals be carried from the land to the ships in the said port, and every of the said keels ought to be of the portage of twenty cauldrons, and according to the same portage the custom is thereof taken to the King's use; there be now certain people which of late have made such keels of the portage of twenty-two or twenty-three cauldrons, whereof the custom hath been taken according to the portage of twenty cauldrons only, in deceit of our lord the King. . .'

In order to put a stop to the shocking leakage of the king's customs, commissioners were appointed to measure the keels and to affix a mark upon them before they were used, and any keel found without the official mark upon it was to be forfeited. No doubt this law had the effect of keeping the keels down to 20 cauldrons. But what of that? The subtlety of the merchants was equal to the occasion; and the cauldron itself grew and grew till it took 12 sacks to make a cauldron while 10 sacks sufficed for a ton. Thus the cauldron came to weigh 4 hundredweight more than the ton.

Why a load swelled out into 5 quarters may perhaps be explained in the same way as the growth of the cauldron. And in like manner we may perhaps explain the divergence between the quantity of standard wheat in a tun measure and that in a ton weight. Such has ever been the result of the conflict between the greed of the king and the subtlety of the merchant.

We may pass lightly over our other measures of capacity. Our bushel is the Norman *bussel*, modern French *boisseau*. The word is doubtless connected with the Italian *bossola*, a little box; and this again is connected with the Old French *boiste*, now written *boîte*. The coomb is a measure of 4 bushels; and various very foolish derivations are suggested in the dictionaries. One considers that the word comes from the Latin *cumulus*; another prefers to trace it to the Greek κυμβος; others again make equally foolish guesses. Surely the word is simply the old British *cwm* a bowl. *Cwm* is still Welsh for a bowl-shaped pool or valley, and it appears in Celto-Saxon dress as coomb and combe in a kindred sense. The origin of the peck I cannot find; it may be the Greek έκτευς, which denoted the same measure. The stoup seems at first to have been a general term for a liquid measure. We have the pint stoup, the mutchkin stoup, and the gill stoup. It is also the name given to the holy water basin in churches. It seems to swell the list of measure-names from Holland; the Dutch form of the word is *stoep*. But it eventually settled down to denote a measure of two quarts or a pottle.

As I happen to have mentioned the Scotch mutchkin, it may be amusing to note *en passant* the droll derivations of the word given by the dictionaries. One says it is from *mete*, a measure, and *han*, a vessel. Another thinks it is simply much-can, in the sense how-much-can. Surely it is merely the diminutive of the Scotch *mutch*, a cap or coif, and means a capful, just as a gallon means a helmetful.

The tun was called by the Anglo-Saxons a *tunna*, but they probably got the word and the thing from the Welsh *tynel*, which is pronounced tunnel and means a barrel. The French *tonneau* is the same word.

The last measure in our tables is called a last, not for that reason, for indeed the two words, though spelt and pronounced alike, have entirely different origins. The last we have now to consider is the Dutch word *last*, meaning a load. Our English load was probably the weight which would be moved about in the ordinary small field-carts of the period, and which held about a ton. The bigger load called a last was the quantity conveyed in the large waggons used for moving heavy goods from place to place over long reaches of country, which seem to have held more than double an ordinary cart load. Thus a last of corn is 10 quarters; but a last of fish is little more than half of that, being only 12 barrels. A last of gunpowder is 24 barrels, and of pitch 14 barrels. A last of wool is 12 sacks, which weigh about a ton and a fifth; while a last of flax is rather more than three-quarters of a ton. Clearly, then, a last is no very precise or universal measure of bulk, and has probably changed as successive legislative attacks on the interests using that measure have been made.

There is only one other measure of capacity of much historic interest, and that is the drachm. But of this I will postpone the discussion till we come to the consideration of the dram weight; and I will reserve for the same place the consideration of the sack, which, though originally a measure, is now regarded as a weight.

And our weights, are they any better in any respect than our bulk measures? To begin with, we have about five distinct systems of weights: troy weight, decimal bullion weight, apothecaries' weight, avoirdupois weight, and metric weight. Several feeble attempts have from time to time been made to bring these systems into harmony, and even to reduce them to a single scale. But one set of persons or another has always succeeded in offering an effectual resistance. In 1824 the Troy pound was taken as the standard, and an attempt was made to bring all our weights into some kind of definite relation with that unit. Since then the avoirdupois pound has been set up in its stead. But, in spite of the report of 1870, the Troy system is still the legal scale for the precious metals; and the absurd system called apothecaries' weight, though abolished by the Medical Acts of 1858 and 1864, was revived by the Act of 1878, and is still lawful in the sale of drugs. The reason for the national toleration of this bewildering muddle is best given in the words of the Royal Commissioners themselves: ' The Troy pound appeared to us to be the ancient weight of this kingdom, having, as we have reason to suppose, existed in the same state in the time of Edward the Confessor; and there are reasons, moreover, to believe that the word " troy " has no reference to any town in France, but rather to the monkish name given to London of Troy Novant, founded on the legend of Brute. Troy weight, therefore, according to this etymology, is in fact London weight. We were induced, moreover, to preserve the Troy weight because all the coinage has uniformly been regulated by it; and all medical prescriptions and formulæ now are, and always have been, estimated by Troy weight under a peculiar sub-division which the College of Physicians have expressed themselves most anxious to preserve.' Doubtless the College of Physicians were anxious to preserve their very peculiar sub-division of Troy weight; otherwise it would have been necessary to invent a brand new set of hieroglyphics, which it would have been difficult to foist upon an awakening public. We have, therefore, been left to wrestle as best we may with the most intolerable tangle of scales the world has ever seen, partly out of deference to an old, exploded, and ridiculous fable about Troy, partly from compliance with the wishes of a set of licensed quacks, and partly because something or other 'always has been done.' Would it be possible to formulate three more childish excuses? As for the apothecaries' foolish mysteries, it is not necessary to suppose that they would find many supporters at the present time. Medical men in our day, with few exceptions, will always, when asked, tell their patients what is being prescribed.

They make no secret about it. It would be quite as easy to write out a prescription in plain English, and in terms of the metric or avoirdupois system, as it is to write it in dog-Latin and in terms of the apothecaries' peculiar sub-division of Troy weight; and it would be much safer and more intelligible. The main object of the medical profession in devising their nomenclature and their scale of measures seems to have been to conceal their own ignorance from their patients.

It would have been unclassical and commonplace to have divided the ounce into any other than twenty-four parts; therefore, the pennyweight had to go to the wall, and the scruple took its place. Then by making twenty grains go to the scruple, the relation between the ounce and the grain was preserved. But the word 'grain' is commonplace. We must coin a classical name for this small measure. Let us call it a minim. Then 60 minims will make a fluid drachm; 8 drachms make an ounce, and 20 ounces shall be a pint; and 8 pints shall be a gallon, which we will always write Cong in order to baffle our patients. It stands for congius, which was nothing like a gallon; but that does not matter. The pound we will write O, the ounce ℥, the drachm ʒ, the scruple ℈, and the minim ♏. Furthermore, we will not use the ordinary Arabic numerals, but the tedious Roman. In order further to thwart the too curious, we will buy our drugs by avoirdupois weight and sell them by our own peculiar system. If then we are careful to write the names of our drugs and materials in Latin and Greek, and to abbreviate even these, we may safely entrust our prescriptions in the hands of the most prying and suspicious. Something like this seems to have been the *rationale* of apothecaries' weight.

Twenty ounces to the pint and 8 pints to the gallon make 160 ounces to the gallon, and this is precisely the number of avoirdupois ounces in an imperial gallon. But the ounce itself is a Troy ounce containing 480 grains. How these facts are reconciled is a question of very slight interest; for in the case of ninety-nine out of a hundred drugs it matters little whether one takes the precise quantity prescribed or double or half. It is hardly necessary to refer to the ineffable twaddle about Brutus and the Trojans, except for the purpose of congratulating ourselves on the spread of education during the last two generations. There might be something—very little, but still something—to be said for the coinage argument if there were exactly five or four sovereigns in a Troy ounce of pure gold or of standard gold; or if a pound sterling contained exactly one hundred grains of pure gold; or if there were any precise relation whatever between the Troy system and any one of our coins. All that can be said is that a silver penny weighed a pennyweight in the times before Edward I., and that a silver pound then weighed a pound. To put this forward as an argument in favour of retaining the Troy system or any other system is of course mere pedantry.

The battle between the Troy and the avoirdupois systems is simply the old fight between the duodecimal and binary scales. The Normans supported the former, but it was always very unpopular with the English; and in the sixteenth century an Act was passed in the twenty-fourth year of Henry VIII., legalising what was called Haver-de-pois. The Act runs thus: 'Beef, pork, mutton and veal, shall be sold by weight called Haver-de-pois. No person shall take for a pound of beef or pork above a halfpenny, nor for a pound of mutton or veal above three farthings, and less in those counties where they be sold for less.' Such was the Englishman's love of the binary scale that this butchers' weight little by little elbowed out Troy weight in every field except that of the precious metals and that of drugs and medicines; the one, be it noted, chiefly a State department, and the other a State-bolstered monopoly. What the expression avoir-du-pois exactly means I confess I do not know. To say, as the dictionaries, do that it means to have weight is perfectly obvious but utterly useless. Certain early English statutes have the phrase, but it is used to signify some ware or other. Thus in 16 Ric. II., c. 1., we have this: 'Every merchant and other of what condition that he be, as well alien as denizen that bringeth wines, flesh, fish, or other manner of victuals, cloth fells or avoirdepois or other ware or merchandises to the City of London or other cities . .' What was this 'avoirdepois'?

. What, again, is the meaning and origin of the Troy pound? Dispensing with the assistance of the Trojans, we find throughout the whole of the early English period that 8 pints make a gallon and that 8 pounds make a gallon. I have said that *pyndan* in Anglo-Saxon means to pen-in or contain. The pynd, then, would signify either the vessel containing or the weight contained; and the word was used indiscriminately as a weight or a measure of bulk. It is quite possible, and indeed probable, that when foreign trade brought the English in contact with the Continental word *pondus*, they would confound it with their own word *pynd*. That this was so is made plain by the fact that the old pynd or pen for stray cattle and other beasts is now spelled pound, though no one pretends that a sheep pound has any connection with the weight or pondus. The pynder was the petty officer of a manor whose duty it was to impound all strange cattle straying upon the common.

> In Wakefield there lives a jolly pynder,
> In Wakefield all on a green.

Mr. Gomme points out in his *Village Community* that in Scotton, in Lincolnshire, we have Pinder's Piece, and in Barrow, Pinder's Thing.

The twelfth part of the Troy pound was classically called *uncia*, in French *once*, and finally in English ounce; and eventually the term 'ounce' seems to have supplanted the old English name for the sixteenth part of the pynd. This may have been owing to the common because convenient practice of weighing things by coins in the good old days when a pennyweight was really the weight of a penny and a pound's worth of silver weighed a pound. Unfortunately for this convenient practice, subtle and crafty persons soon discovered that clipping the silver penny was a lucrative kind of business; and an early law tells us that a pennyweight is the weight of a silver penny 'round and without any clipping.'

Standard silver in those days seems to have been fifteen-sixteenths fine, and the silver pound originally consisted of 256 pennies. It is clear that 240 real minted pennies would therefore weigh 256 pennyweights. Hence it was said that 240 pennies made a pound, and the confusion between the pound weight and pound sterling seems to have been taken advantage of by moneyers, the royal mint and tax-gathering classes. The old pynd contained 8,192 grains; take away one sixteenth to represent the alloy, and we have 7,680 grains left; this is exactly the weight of Edward III.'s pound. A pound came to contain, not 8,192 grains of pure silver, but 7,680 grains, and this actually came to be regarded as the real number of grains, not only in a £ but also in a lb. Thus, 240 pennies made a pound both in coinage and in weight, and they were reckoned either in scores or in dozens. A dozen pennies made a shilling, and a score made an ounce. Hence, a score of shillings made a pound, or a dozen ounces made a pound. Therefore, a shilling was three-fifths of an ounce, and an ounce was one-and-eightpence. Our present shilling contains of pure silver three and a third pennyweights, and it is worth rather more than double that, by reason of the gold which it is taken to represent, and for which it can be exchanged at any time. These ratios give no idea of the relative values of the two shillings, owing to the enormous depreciation which has taken place in the value of silver with respect to things in general since that period. In order to realise this, we must compare the purchasing power of a given weight of silver then and now. 'Thus,' says Eccleston, ' the price of labour seems to have varied from about three-farthings to a penny a day, with victuals. The prices of grain varied excessively, even at different periods of the same year; wheat perhaps generally averaged four shillings the quarter, though in scarce years it sometimes rose to a pound. In 1185 sheep were rated at about fivepence-halfpenny each, hogs at a shilling, cows at about four shillings and sixpence, and breeding mares at less than three shillings.' We must pay no attention to the fancy prices paid then, as now, for the luxuries of the rich, such as jewels and war-horses; but we are told that two arches of London bridge cost only twenty-five pounds. If one of our 1894 shillings had fallen out of the sky into the hands of one of William the Conqueror's retainers, it would have been worth less than threepence-halfpenny, whereas to a modern silversmith it is worth over fivepence. On the other hand, the Norman soldier would have been able to purchase with it more than half a sheep, whereas our silversmith would be able to buy only about a pound of mutton.

In the north country there is a kind of gruel called 'drummuck.' In the eastern counties one who is sodden with drink is said to 'drumble.' There was a great noise made in the days of the Plantagenets about the thrums or ends and rubbish of weavers' threads, under cover of which good thread was moved abroad without paying the high duty then customary. The lees or dregs of liquors seem to have been called 'thrums' also. The Dutch word for lees and sweepings of threads is *drom*. The Danish for dregs is *drank*; and we also talk about the drains left in a cup: this is the Saxon *drehnigean*, to drain. The Anglo-Saxons had a coin called a thrimsa, supposed to have weighed about $67\frac{1}{2}$ grains of silver, perhaps 64. On the whole, I cannot help suspecting that there was an old small measure for drugs or dregs, and ardent spirits known to the English by some such name as dram. With their usual adaptiveness they would be quite ready to adopt the Greek word $\delta\rho\alpha\chi\mu\eta$ as soon as it was introduced by the doctors. We know how readily Greek and Latin spellings were adopted for plain English words at the time of the Renaissance. The Flem was taken as the type of dullness, heaviness, and apathy, just as the Frank has always been the type of gaiety and light-heartedness. But Flem and Flemish were changed to phlegm and phlegmatic, as though they denoted something fiery and brisk. Again, the English rime came to be spelt rhyme, as though derived from some Greek word like $\rho\upsilon\theta\mu\text{o}\varsigma$. To this day rodomontade is always spelt rhodomontade, as though it had something to do with the Greek for red, like rhododendron. The reverse process is unusual; and I cannot see how such an out-of-the-way weight as the Greek drachm can have become so popular as to work its way, not only into the apothecaries' table, but also into the butcher's bill; and, furthermore, to appear in English garb as dram. Besides, it was a colloquial expression. People used to say of a fellow that he had not a dram of wit. Like the pound, I am disposed to accord to the dram two distinct origins, one Anglo-Saxon and the other foreign. The English dram appears to have been the sixteenth part of the sixteenth part of a pound; whereas in Troy weight, as peculiarly subdivided by the apothecaries, it is the eighth part of the uncia. What the sixteenth part of a pound was called before it came out in a Roman name I do not know.

And what about the grain, our real or imaginary unit of weight? An Act of Henry VII., purporting to revive the old Saxon pound under the name of Troy-pound, divides it into 12 ounces of 20 sterlings each, each sterling to be of the weight of 32 corns of wheat 'that grew in the midst of the ear, according to the old laws of this land.' And yet it is said that this pound of Troy weight turned out to be one-sixteenth heavier than the pound in vogue at the time. According to later Acts, the pennyweight contained only 24 grains, and consequently a grain troy became a heavier weight than the standard grain of wheat. Whatever the grain was, its weight must have been considerably altered. It may have been an absurd pretence that the pennyweight was actually based on real grains of wheat from the middle of the ear, just as it is an absurd pretence that the inch was based on the length of 3 barleycorns placed end to end, and just as it is an absurd pretence that our modern yard is based on the length of the pendulum swinging seconds, and that our modern pound was derived from the weight of a cubic inch of water at a temperature of 62° Fahr. weighed in air. In all these cases it seems probable that the standard had been already fixed, and the pretence of deriving it from something in nature was made merely to give colour to the contention that such standard unit had a meaning; whereas neither the pound nor the yard had any more meaning than if the former had been the first stone picked up in the road, and the latter had been the first stick torn from a tree. It is fair to add that the Weights and Measures Act, 1878, discards all these pretences, and in effect says boldly: 'This bronze rod is a yard because we say so, and for no other reason whatever; and this platinum lump is a pound for the like reason. We certainly do not profess to have divided a cubic inch of distilled water into 252,458 parts, and to have taken seven millions of such small quantities of water (thousandths of a grain) and so built up our pound; firstly, because we could not divide a cubic inch of water into 252,458 equal parts to save our lives; and secondly, because if we could it would be a useless task, for a cubic inch is no less an arbitrary quantity than this lump of platinum. Neither do we profess to have derived our yard from the length of the seconds pendulum; firstly, because we tried to do this for about twenty years, and failed; and secondly, because the conditions imposed upon the pendulum are in all respects as arbitrary as this bronze rod.' Similarly, we may regard the grain base of the old English system as a simple myth. King Henry, or someone under him, put a silver penny in one scale and tried how many grains of wheat it took to balance it in the other, and then gave it out that that was how the pennyweight was got at. And when, later on, the duodecimal craze came in again, it was decided that 24 grains *ought* to weigh a penny; and if on trial it turned out that they did not—why, then, so much the worse for the grains; the penny would remain just where it was. The nearest binary figure which would approximate to the average number of wheat-corns in a pennyweight would be 32, and the nearest duodecimal figure would be 24. Hence we find that there were at first 32, and afterwards 24 grains in a pennyweight, although there is good reason to believe that the ounce itself had not then varied at all.

About sixty years ago Sir William Betham published some curious observations on ancient Celtic ring-money. According to him, specimens of this primitive currency, both of gold, silver, and bronze, have been found in great numbers in Ireland. Sometimes the form is that of a complete ring, sometimes that of a wire or bar bent round till the two ends are brought near together. In some cases these ends are armed with flattened knobs. Sometimes several rings are joined together at the circumferences; sometimes they are linked into one another in the form of a chain. But what I wish to draw particular attention to is that, upon being weighed, by far the greater number of them appeared to be exact multiples of a certain standard unit.

The smallest gold ring which Sir William had seen, weighed exactly half a pennyweight; another contained exactly 3 times this quantity; another, 5 times; another, 10; another, 16; another, 22; another, 480; and another, 534. Sir William says that this is so, not only in the case of the gold rings, but also of the silver and of the bronze. All of them, with very few exceptions, which may easily be accounted for on the supposition of partial waste or other injury, weigh each a certain number of half-pennyweights. The smallest specimens even of the bronze ring-money are quite as accurately balanced as those of the gold and silver; and amongst the bronze specimens he states that, after having weighed a great many, he never found a single exception to their divisibility into so many half-pennyweights.

It would thus appear that the pennyweight has not varied much from the very earliest time to the present. Whether Sir William is right in regarding this ring-money as Celtic, is a question for the antiquary; but having regard to the fact that gold and silver ring-money appears in the fresco paintings in the tombs of Egypt, one would be justified in surmising that the specimens found in Ireland are more likely to have been brought over by the Phœnicians than to have formed part of a native currency. Our present interest in the matter rests on the fact that at this early date the unit of weight was precisely 16 old grains. What the weight of the old grain was as a standard unit we shall endeavour to find out in the next chapter.

It is now time to overhaul our table or tables of weight for the hundredth time—this once without the assistance of the Quacks, the Mint, or the Trojans. Taking the avoirdupois scale we have the following:—

16 drams . =1 ounce
16 ounces . =1 pound
8 pounds . =1 butchers' stone
14 pounds . =1 stone
2 stones . =1 quarter
4 quarters . =1 hundredweight
20 hundredweight =1 ton

Let us assume, as we did with the gallon table, that originally these weights were arranged on the binary system, and let us substitute the nearest binary numbers for those in the table. This gives us the ounce of 16 drams, the pound of 16 ounces, the little stone of 8 pounds, the big stone of 16 pounds, the quarter-weight of 2 big stones, the weight of 4 quarters, and the ton of 16 weights.

These names are hardly convenient. But if we look about the country we shall find certain other weights used for various commodities, which may help to throw some light on the subject. Wool and cheese are to this day sold by the clove and wey. Henry VI. having abolished the auncel, a most deceitful instrument for weighing, the new apparatus which took its place appears to have greatly perplexed the vendors of cheese. They could not understand the new system, and it became necessary to explain it to them by Act of Parliament. This was done by 8 Hen. VI. c. 8: 'Whereas it hath been of old time accustomed in all the counties of England that all the cheeses which ought to be sold by the wey should be weighed by the auncel, and because that at the last parliament holden at Westminster it was ordained that the said auncel, in respect of the great deceit of the same, should be destroyed, and other weights couching should be in this behalf ordained; and it is so that the poor people of the realm be greatly deceived by the said weights couching, for that they know not how many pounds the wey of cheese doth contain by the said weights couching : therefore, to the intent that the said poor people shall not be in this behalf deceived as they have been sithence the said last parliament, it is ordained that the weight of the wey of cheese may contain 32 cloves, that is to say, every clove 7 pounds by the said weights laying.'

And what may a clove be? Consider first that we have already a lawful stone of 16 pounds—at that time already reduced to 14— and that merchants and dealers were provided with such lawful stones. For the half-stone it would be a simple matter to cleave the stone in twain, whereupon each half would be a cloven stone or clove stone. This gives us the lesser stone used by the butchers, and it also gives us the clove used to this day by the wool merchant. And 32 of these old cloves of 8 pounds each give us exactly the binary figure 256, and to this day the weight of a wey of cheese in Suffolk is 256 pounds. The word wey is only another form of the word weight. A modern wey of wool has been reduced by the subtlety and craft of merchants to 26 cloves of 7 pounds to the clove, making it only 182 pounds.

The modern wool table, though rapidly dying out, is this :—

7 pounds .	= 1 clove
2 cloves .	= 1 stone
2 stones .	= 1 tod
6 tods	. = 1 wey
8 tods	. = 1 pack
2 weys .	. = 1 sack
12 sacks .	. = 1 last

It seems feasible to work these two tables into one, each furnishing missing links for the other. Then, substituting the proper binary numbers, we get something like this :—

16 drams .	= 1 ounce
16 ounces	= 1 pound
8 pounds	= 1 clove or cloven stone
2 cloves .	= 1 stone
2 stones .	= 1 tod or quarter-wey
4 tods	= 1 wey (hundredwey)
2 weys .	. = 1 pack (packwey)
8 packs .	. = 1 ton
2 tons .	*. = 1 last of 4,096 pounds

There appear to have been two weys or weighs, one of 128 pounds, and the other of double that weight. There is no cleaving to be done here. How, then, shall we account for it? A glance at the illustration on the following page will reveal the solution.

'The little trade which existed between one part of the kingdom and the other was carried on by means of pack-horses, along roads little better than bridle-paths. These horses travelled in lines, with the bales or panniers strapped across their backs. The foremost horse bore a bell or a collar of bells, and was hence called the bell-horse. He was selected because of his sagacity; and, by the tinklings of the bells he carried, the movements of his followers were regulated. The bells also gave notice of the approach of the convoy to those who might be advancing from the opposite direction. This was a

matter of some importance, as in many parts of the path there was not room for two loaded horses to pass each other, and quarrels between the drivers of the pack-horse trains were frequent as to which of the meeting convoys was to pass down into the dirt and allow the other to pass along the bridle-way. The pack-horses not only carried merchandise, but passengers, and at certain times scholars proceeding to and from Oxford and Cambridge. When Smollett travelled from Glasgow to London he rode partly on pack-horses, partly by waggon, and partly on foot; and the adventures which he described as having befallen Roderick Random are supposed to have been drawn in a great measure from his own experiences during the journey.'

This description and the illustration from Dr. Smiles's charming *Lives of Engineers*[1] show us at once that the big wey or packwey, such as a horse could carry on his back over long reaches of bad road, was divided into two small weys, one to be set on either side of him. The two weys together made what came to be called a pack, and a half-pack came to take the name of wey. But in the case of some articles that could not, like wool, be stowed at will in two separate panniers, the old wey would retain its meaning of a full packwey, as it has done in the case of cheese.

How did the wey of 128 pounds come to be called a hundredweight? The Royal Commissioners of 1819 tell us that it originally consisted of 100 pounds, and that it eventually increased to 112. Not only is there no evidence of this, but it is opposed to all the probabilities; and, moreover, there is a simple explanation. As we see from the Harleian MSS., the wheat was carried from the granary to the fields in sacks. These sacks were made of uniform size and shape, and all other sacks were forbidden by law. If you take a sack made to contain 128 pounds of wheat and fill it with coal, you will find that it will hold just about 200 pounds. This was and is the sack of coals, and the half of it is rightly called a hundredweight. It was necessary to give this weight or wey a special name in order to distinguish it from other weys. We never speak of a hundredweight of wheat. We have seen that the wey of wool is 182 pounds, and that the wey of cheese is still in some places 256 pounds. Indeed, the word is a generic term meaning any definite weight whatever. For example, in imperial dry measure a wey is 5 quarters, or 40 bushels, or 2,560 pounds, that is, just ten times the weight of a Suffolk wey of cheese. It would, therefore, be absurd to suppose that these 3 weys ever had anything in common but the name. The like is true of the term quarter, which is used to signify the fourth part of several different weights

[1] *Lives of Engineers*, by Samuel Smiles; John Murray, London, 1861.

and measures. It means the quarter of a hundredweight, which is 28 pounds; and it means 8 bushels, which is 512 pounds; and again, in the form of quart, it appears as the quarter of a gallon.

The ton was originally the weight of a tun or two pipes of standard wheat. But such has been the effect of legislative tamperings with weights and measures that a bushel of corn not only weighs less than a bushel of coals, but also measures less, which is ridiculous. Instead of saying the bushel of coal shall be of such a measure, the State said the bushel shall be of such a weight—in other words, it shall not be a bushel. A bushel of wheat contains 2,218 cubic inches, whereas the coal bushel contains 2,815 cubic inches, a difference of nearly 600 cubic inches. Like discrepancies occur in all our special systems.

Ordinary hard stone 4 inches cube weighs 8 pounds; therefore the original stone was probably a real stone measuring 8 inches long, 4 wide, and 4 thick, and the lesser or clove stone was a stone of 4 inches cube. Coal weighs one and a quarter time as much as water; hence a sack which would contain 128 pounds of wheat (that is to say, of good wheat of 64 pounds to the bushel) would contain 160 pounds of water and 200 pounds of coal. Hence, as we have seen, a hundredweight was the weight of coal contained in a sack made to contain 64 pounds of wheat. In other words, a hundredweight is the weight of a bushel of coal.

The tod appears to have been originally a weight of 32 pounds. Its name is British, and, like the other three British measures, it is used to denote the double of one of the Saxon standards. Thus the ton is the double pipe, the coomb is the double boll or wey, the tod is the double peck or stone, and the potel is the double can or quart. This disposes of the names of our ordinary weight measures.

CHAPTER VI

THE OLD PIPE SCALE

IT is customary to speak of the pound as the unit of English weight. The laws of Plantagenet times appear to regard the grain as the unit from which all other weights were derived. But we have seen that possibly the standard grain was a myth. What the original unit was we have now to seek. And I think I may say in advance that the result of our search will be to show that the unit from which all our weights, as well as our measures of capacity, were derived is none other than the yard itself. We know with what jealous care this nation has ever preserved its semi-sacred standard. As the Royal Commissioners of 1869 affirm, not without pride, our yard has not varied the sixteenth of an inch since the days of Edward the Confessor, and even earlier. We will now go to the British Museum, where we shall examine an old drawing in one of the Harleian MSS. On the top left-hand corner you will see two measures, one of them a large tub or butt, and the other a small one, in the hands of a man who is measuring out the corn. Examine the large vessel carefully. It is well known that early artists were scrupulously careful in getting the correct scale of all standard measures depicted by them. An inch or two in a man's leg was a trifle, but if a mete-yard or a congius had to be introduced into the picture, it was correctly and almost microscopically drawn to scale. Taking the measuring man to be of average height, it is pretty obvious that the tub or butt close by him is intended to measure one yard high and one yard in diameter. Remember that these people were an agricultural people, and remember further that there was a tradition long prior to the Conquest that all our measures were or had been based on the grain of wheat. Whether true or not, the pretence is still more to our purpose. It was said that 3 grains of barley went to the inch. Efforts were even made, by changing the kind of grain, to fit the fable to the fact. The children in the Board School are taught to this day that 3 barleycorns make an inch, which they do not. Again, our fathers were taught that a pennyweight was the weight of 32 grains of wheat from the midst of the ear; and, when that was found to be by no means invariably true, the number was altered to 24, in

order that the glamour of the duodecimal system might give the fable some support.

One thing is certain, and that is, that the unit of bulk and the unit of weight were correlated, with wheat for the medium. Otherwise the pound could not have been the weight of a pint. That is to say, a pint of water weighs a good deal more than a pound; whereas a pint of good wheat weighs just about a pound. Let us be as exact as possible in our calculations. And, first of all, what quality of wheat was in all probability used as standard wheat for the purpose of connecting and adjusting weights and measures. Let us assume that the relation between the avoirdupois pound and the modern pint is just what it was originally—that is to say, that the pound weight is the weight of a pint of standard wheat. Now, our modern pint is $\frac{1}{32}$ less than the pint of the pipe scale. On these simple data let us find the number of Gothic pounds in a yard cylinder of standard wheat. A cubic inch of distilled water under ordinary conditions weighs exactly 252·458 Troy grains. We have this on the authority not only of scientific men, but also of the infallible State official—nay, of the Statute Book itself. Therefore, a cubic yard of water weighs 1,682·6686 avoirdupois pounds. And a cubic yard of wheat at ·8 comes out 1,346 within a couple of ounces. Take away a thirty-second part of this in order to allow for the difference between the old pint and the new. This leaves 1,304 old pounds for the weight of a cubic yard of standard wheat. But our forefathers did not use cubic measures. They used cylinders like those in the old picture before us. We must therefore multiply by ·7854, and the product will be the number of old pounds or pynds in the yard cylinder of standard wheat. The answer is 1,024.

Surely this figure is familiar to us. It is our old friend the binary figure 2^{10}. This is something more than a coincidence. The yard cylinder and the old pipe turn out to be one and the same thing to an ounce, when measured in wheat of a specific gravity of ·8. At any rate, I for one have not the smallest doubt that the original Gothic standard bulk measure was a yard cylinder, and that it was called a pipe, and that we see it as it were in the flesh in the ancient picture before our eyes. To me it seems also certain that the standard unit of weight was the weight of this measure of standard wheat, or four-fifths of the same measure of water. Whether the name of this weight was pipe-weight, just as the pynd-weight was also called a pynd, I do not know. The double of this pipe-weight was called a ton, and the bulk measure was called a tun. Originally both tons were spelt the same, but they have since been differentiated.

If we construct another cylinder half the height and half the diameter of the pipe, we see that it contains two bushels of standard wheat. It seems to have been formerly called a boll, and its weight is the ancestor of our modern hundredweight. If we now construct a span or quarter-yard cylinder, we have before us the old peck, the weight of which was the stone. This is probably the small measure which we see in the measurer's hands.

Constructing another cylinder one-eighth of a yard in diameter, inside measurement, and an eighth of a yard deep, we have before us the Saxon can, our quart pot. And, again halving the dimensions, we have before us the gill measure. This is the nail cylinder. Let us compare the heights and diameters of these hypothetical measures with the actual measures in the hands of the Board of Trade:—

		Inches	
Yard-cylinder or pipe	36	35·61836	Sixteen-bushel
Half-yard or boll	18	17·80928	Two-bushel
Quarter-yard or peck	9	8·90464	Peck
Eighth-yard or can	4·5	4·45232	Quart-pot
Sixteenth-yard or gill	2·25	2·22616	Gill

Comparing these, we find that the difference between the old pipe and the modern sixteen-bushel imperial is less than two-fifths of an inch. The difference between the height of the quarter-yard cylinder and our modern peck is less than one-tenth of an inch. Between the old can and the modern quart pot the difference in height and diameter is less than a twentieth of an inch. And between the nail cylinder and the modern gill the difference is less than a fortieth of an inch.

Surely we are forced to the conclusion that these were the original measures of capacity in this country, and that they were all based on the yard, the form of the vessel being the cylinder, and the medium by which the weight measures were connected with the bulk-measures being standard wheat. We must remember that the bushel or half-strike would be a little larger than our present bushel—about one thirty-second larger. Having regard to the Anglo-Saxon predilection for binary figures, they would of course adopt the standard of 64 pounds to the bushel. Indeed, they had no choice, because the pound was the weight of the pint, and as 64 pints go to a bushel, a bushel must necessarily weigh 64 pounds. We now see that the can or quart pot weighed exactly 2 pounds, the peck weighed exactly a stone of 16 pounds, and the strike or boll weighed exactly a wey of 128 pounds.

Judging from the names of the measures, one would suppose that the early constructors of this system were satisfied with bulk measures rising by eights, and all represented by cylinders of which the height was equal to the diameter. Later on it would appear that the doubles and halves of these measures were brought into the system. It may be a coincidence, but it is a fact, that all four of the double-cylinders received British names. The double pipe was the tun; the double boll was the coomb; the double peck was the tod; and the double can was the potel. Then the half-cylinders seem for the most part to have received Norman names. Thus the half-pipe is the *quartier* or quarter; the half-boll is the *boisseau* or bushel, while the half-peck is the gallon; an exception occurring in the case of the half-can, which is the famous pint or pound. The reason why a half-cylinder was chosen for this important measure is probably to be found in the history of the currency.

The cylinders in this fine old system went down from the yard in height to the sixteenth of a nail, or the two hundred and fifty-sixth part of a yard. The double of this microscopic cylinder is the mythical grain of standard wheat from the midst of the ear, and it is a very remarkable fact that these are actually the dimensions of a well-grown wheat-corn, which will be found on measurement to be about a sixteenth of a nail in thickness and about double that in length. This is the one circumstance which would lead one to suppose that the inventors of this system did really build it up from the actual grain. For my own part I am disposed to believe that the system was based upon the yard, which was originally some particular staff or wand, and that the extraordinary relation between the bulk of the wheat-corn, dry and from the midst

F

of the ear, is merely a remarkable coincidence. Others can believe what they choose, for the beauty of the system is hardly affected by the truth or untruth of the hypothesis. Mr. Norris's experiments on the actual weight of wheat grains are untrustworthy, and evidently strained to make the figures fit the fact. He says that $22\frac{1}{2}$ average grains of good dry wheat weigh exactly a pennyweight. I have tried it and find it untrue. What the particular staff or wand originally was is a question of the deepest interest. I shall show that in all probability it was about $1\frac{1}{2}$ per cent. longer than our present yard, and only one-thousandth longer than the corresponding Greek measure—the διπηχυς.

We know that the Gothic pound contained 8,192 Gothic grains ($32 \times 16 \times 16$). And we know that our present avoirdupois pound contains 7933·3 such grains, or 7,000 modern Troy grains. The difference between the two pounds is therefore just under $\frac{1}{32}$ of the larger, or $\frac{1}{31}$ of the smaller. But we also know that the old pound was the weight of a Gothic pint of wheat of specific gravity ·8. And our modern avoirdupois pound of such standard wheat exactly fills our modern pint. We are justified in concluding that the relation between the pound and the pint has been preserved.

What, then, was the size of the Gothic pint in cubic inches? Our modern pint contains 34·66 cubic inches. We have, therefore, to add $\frac{1}{31}$ of this to make the Gothic pint, $34·66+1·12=35·78$. This is a very remarkable figure. This number of cubic inches is ·000767 of a yard. Now, 8 pints made a gallon, 8 gallons a bushel, 8 bushels a quarter, and 2 quarters a pipe. Multiplying ·000767 by 1,024, we get pipe=·7854 of a cubic yard. *And this is a yard cylinder.* There can be no doubt whatever that the Gothic unit of bulk-measure was a cylinder 1 yard high and 1 yard in diameter. And as we proceed with our inquiries the evidence accumulates. The above is only one out of half-a-dozen ways in which the theory can be shown to be more than probable. It might have been guessed from a comparison of the following figures, showing the height and diameter of our modern English

THE OLD PIPE SCALE

measures expressed in cylinders, compared with corresponding cylinders of the yard series.

		inches	inches	
Cylinder yard	.	. 36	35·61856	Double-quarter
,,	half-yard .	. 18	17·80928	Strike
,,	quarter-yard	9	8·90464	Peck
,,	eighth of yard .	4·5	4·45232	Quart
,,	sixteenth of yard .	2·25	2·22616	Gill

Thus it appears that our double-quarter (formerly the pipe) in the form of a cylinder is a yard high within about a third of an inch, and that all the other measures of bulk similarly approximate within minute fractions to the cylinders based on the binary sub-divisions of the yard.

All the above cylinders are of the form in which the height equals the diameter. It will at once occur to the reader that this list does not include some of the most important English measures, such as the gallon, and bushel, and pint. We also know that our old English grain-measures rose in a binary progression, thus:—

2 pints	=	1 quart	2 bushels	=	1 strike
2 quarts	=	1 stoup	2 strikes	=	1 coomb
2 stoups	=	1 gallon	2 coombs	=	1 quarter
2 gallons	=	1 peck	2 quarters	=	1 pipe
2 pecks	=	1 tod	2 pipes	=	1 tun
2 tods	=	1 bushel			

and so on. Now what was the form of these other cylinders?

If we double the height of a cylinder, leaving the diameter the same, it is obvious that we double the bulk, but if we leave the height unaltered and double the diameter, we quadruple the bulk. Hence, we obtain a binary series of cylinders as follows:—

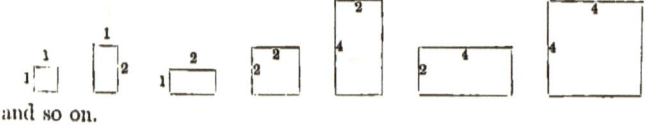

and so on.

Although the Romans when the quadrantal was established, and more recently the French when the Metric System was introduced, adopted cubical bulk measures, it is certain that the oldest and universal practice was to measure bulks, whether of grain or liquids, by means of cylinders. To begin with, cylinders are far easier to make than cubic measures. Secondly, the field measures of all countries have been cylinders as a fact. Thirdly, the very word 'pipe' means a cylinder, and so does the Roman word *culeus*, which was (as I hope to show) originally the same measure. Κολεος was the sheath of a sword or the bark of a tree, and it is highly probable that the first pipe measure was the bark of some tree a yard in diameter (inside measurement) and one yard long. The core would be taken out after drying, and, after cutting off a section an inch or so in thickness for the bottom of the tub, the vessel would serve as a convenient and practically accurate measure for corn.

We are now in a position to sketch out the skeleton of the old pipe scale. It is a remarkable fact that the *form* of the several cylinders should have persisted to this day. We know that a pound weight was the weight of a pint of something. Indeed, both the bulk and the weight measure were called by the same name, a name I propose to adopt in describing the old Gothic weight, so as to distinguish it from the other pounds to which we may have to refer. That name is pynd.

We have seen that the pynd contained 8,192 Gothic grains, whatever they may have been: namely, $32 \times 16 \times 16$. I shall show as we proceed that the Gothic grain was exactly $\frac{13}{16}$ of the Tower or Colonia grain used by our moneyers up to the time of Henry VIII., and that it was $\frac{15}{16}$ of our present Troy grain. I shall ask to be allowed to anticipate this in order to proceed with the reconstruction of the pipe scale of weights and measures, promising faithfully to prove both statements up to the hilt later on.

It must not be supposed that the names given to these bulks and weights are those used by our Gothic ancestors before the Saxon invasion, though most of them probably are. But others are borrowed from foreign sources, and others again are of recent origin. Thus our quart was formerly known as a can or *kan*, and the word 'strike' seems to be derived from the custom of striking the bulk measure with a round stick or strickle, so as to make it level at the top instead of heaped. Yet, though the names may have changed, the measures themselves are almost precisely what they were a thousand, and perhaps (as we shall see) five thousand, years ago.

We have now to ascertain the medium whereby bulks and weights were connected. Taking the yard cylinder as ·785398, and the weight of a cubic inch of pure water as 252·468 Troy grains, we find that the old pynd contained 9034·13288 Troy grains. Therefore, it contained 10,240 Gothic grains of pure water. But we know that it weighed 8,192 Gothic grains. Therefore, the medium used must have been something having a specific gravity of $\frac{8192}{10240}$, which is exactly ·8. It is hardly necessary to waste time in showing that the medium used was wheat. Firstly, we are told so in all our laws and statutes relating to weights and measures from the days of Edward the Confessor downwards. Secondly, it is generally admitted that grain was used as the universal medium among agricultural people. Then the word 'grain' itself is significant. Finally, the specific gravity of good wheat, dried, 'from the midst of the ear,' as by law directed, is 64 pounds to the bushel, or ·8, or thereabouts. Hence, there can be no doubt whatever that the Gothic weights were the weights of certain bulk measures of standard wheat, such standard wheat being of a specific gravity of ·8—that is, four-fifths of the weight of an equal volume of water.

Working out the problem, not in round figures and approximations but to five rows of decimals, we find that the old pynd weighed 7,228 Troy grains, *within a grain*. That is to say, by following one line of calculation it comes out 7,227·3063, and by following another line it comes out 7228·24—difference less than a grain in the pound. There are those who pretend that the Gothic standard wheat was 62 pounds to the bushel, but it is clear they have a theory to bolster up. Besides, no community making use of the beautiful pipe scale with its similar cylinders based on the yard and its binary sub-multiples, and rising by eights, would establish any other standard than ·8. Thus a gallon of water weighs a gallon and a quart of wheat, and so on all down the scale. The experiments made last century with a view to ascertain the actual weight of a grain of wheat were conducted without any regard to the directions expressly laid down by statute. I have myself made numerous observations on the average weight of good sound wheat well dried and taken from the midst of the ear, and I find that so far from being less than 64 pounds to the bushel, it is, if anything, rather heavier. It is hardly necessary to point out that a grain of wheat taken by itself is *heavier* than water. It is only by reason of its shape that a mass of wheat is lighter than an equal volume of water, owing to the fact that we are forced to measure the interstices between the grains.

Anyhow, the evidence is overwhelming that the wheat selected as the standard by the Teutonic people weighed 64 pounds to the bushel or ·8 of water; this will become more manifest as we proceed. I shall show that the Romans used the same standard before the adoption of the quadrantal, and that the Greeks used a still heavier wheat having a specific gravity of ·833 or $\frac{5}{6}$. All the Gothic measures of bulk and weight may now be expressed in grains by the figures 2^0, 2^1, 2^2, 2^3, 2^4 . . . 2^{21}, 2^{22}, 2^{23}, 2^{24}.

Let me re-state the argument :—

If the yard has remained unaltered; if 64 to 80 was the chosen ratio of standard wheat to water; if 8,192 was the weight of a pynd of standard wheat in Gothic grains; then the weight of a pynd of water must have been 10,240 Gothic grains. But the weight of a pynd of water in modern grains is 9,034 Troy grains. Therefore, the ratio of the Gothic grain to the Troy grain is 70·578 to 80, or just $\frac{14}{16}$. Therefore, one pound of 7,000 Troy grains weighs 7,934·5 old grains very nearly. Subtracting this from 8,192, the weight of the pynd, leaves 256·5; and 256 is $\frac{1}{32}$ part of 8,192; therefore, the old pynd weighed $\frac{1}{32}$ more than our modern avoirdupois pound.

Now the old pynd or hand-cylinder halved measured 35·78 cubic

THE OLD PIPE SCALE

Diameter	Form	Bulk	Weight	Gothic grains		Pynds
Yard	Double-pipe	Tun	Ton	2^{24}	16,774,016	2,048
	Pipe	Butt	—	2^{23}	8,387,008	1,024
	Half-pipe	Quarter	—	2^{22}	4,193,504	512
Half-yard or cubit	Double-boll	Coombe	—	2^{21}	2,096,752	256
	Boll	Strike	Wey	2^{20}	1,048,376	128
	Half-boll	Bushel	—	2^{19}	524,288	64
Quarter-yd. or span	Double-peck	Tod	Qrtr-wey	2^{18}	262,144	32
	Peck	Peck	Stone	2^{17}	131,072	16
	Half-peck	Gallon	Clove	2^{16}	65,536	8
Eighth yard or hand	Double-kan	Stoup	—	2^{15}	32,768	4
	Kan	Quart	—	2^{14}	16,384	2
	Half-kan	Pint	Pound	2^{13}	8,192	1
Sixteenth or nail	Double-gill	Half-pint	Mark	2^{12}	4,096	
	Gill	Gill	—	2^{11}	2,048	
	Half-gill	—	—	2^{10}	1,024	
32nd or dactyl	Double-ligule	—	Ounce	2^{9}	512	
	Ligule	—	—	2^{8}	256	
	Half-ligule	—	Skilling	2^{7}	128	
64th yard	Double-dram	Fl. drachm	—	2^{6}	64	
	Dram	—	Penny-wt	2^{5}	32	
	Half-dram	—	Sceatta	2^{4}	16	
128th yard	Double-carat	—	—	2^{3}	8	
	Carat	—	Styca	2^{2}	4	
	Half-carat	—	—	2^{1}	2	
256th yard	Grain	—	Grain	2^{0}	1	

inches, and our modern pint measures 34·66 cubic inches ; the difference is 1·12 cubic inches, which is $\frac{1}{30}$. Hence, pynd was to pound in weight as pynd was to pint in measure.

The conclusion to be drawn from these figures is either that all our weights and measures have lost just $\frac{1}{30}$ since Anglo-Saxon times, or that our yard has grown about a third of an inch in length, or, to be more accurate, ·33144 inch. The only other explanation would be that the selected ratio between the standard wheat and water was 62.

To this there are three formidable objections; one is that good dry wheat corns from the middle of the ear do as a fact weigh fully eight-tenths of an equal bulk of water ; another is that the Anglo-Saxons, or whoever constructed their system of measures, were so thoroughly imbued with a passion for the binary scale that they would be certain to select 64 to 80, that is ·8, or some other simple binary ratio ; and finally, the hypothesis does *not* explain the equal loss of both weights and measures. For my part I have not the slightest doubt that they chose 64 to 80, or a gallon to a gallon and a quart. Thus the pynd measure of *water* would weigh a pynd and a quarter.

How, on this hypothesis, are we to account for the fact that the pre-Norman silver pennies hitherto discovered seem to weigh on the average about $22\frac{1}{4}$ Troy grains? If they do, then they must have weighed $25\frac{1}{2}$ Gothic grains; whereas all early laws distinctly state that they weighed 32 grains—that is, Gothic grains. The only explanation I can think of is this: the sixty-fourth part of a pynd was called a shilling. It weighed therefore 128 grains. Now Athelstan states that 4 pence make a shilling. This is distinctly implied in one of his extant laws. But Canute states quite as distinctly that 5 pennies go to the shilling. Ælfric the grammarian says the same. There is no mistake about this. What are we to make of this discrepancy? If 4 pence went to a shilling, each penny would weigh 32 grains; but if 5 pence went to a shilling, then each penny would weigh $25\frac{3}{5}$ grains, and this is just about the weight of those hitherto discovered, expressed in old grains. I conclude, therefore, that the pennies which we have found are the lesser pennies, of which five went to the shilling. Whether any pennies weighing 28¼ Troy grains have yet been found I do not know. But this is what a penny of 32 old grains would weigh in modern Troy grains. If this is the correct solution, it would seem to settle the question whether the yard has grown bigger or the popular pound has grown less. And if at any time any Saxon pence should be found weighing 28¼ Troy grains, or, to be more accurate, averaging 28·23 Troy grains, there could no longer be any doubt on the point.

Anyhow, it is a singular confirmation of the theory, that the Saxon pennies hitherto found should weigh just about 25¾ old grains, which is exactly a fifth of 128 grains, or an A.S. shilling, a ratio which precisely accords with the statements both of Canute and of Ælfric.

4 gills or nail cylinders = pynd = 2^{13} Gothic grains = 8,192, and 32 such grains made a Gothic penny. The Saxon moneyers, however, adopted the Colonia pound, which was three-quarters of the old pynd, and also the Roman sub-division. Hence the penny contained 24 Colonia or Tower or moneyer's grains. When Edward and his successors say that a penny is 32 grains of wheat dry and from the midst of the ear, they must either be confounding the current penny with the Gothic or market pennyweight, or else the grain of the day must have been temporarily degraded to smaller proportions than has ever existed elsewhere, or than is in accordance with facts. I venture to think that the grain of that day was almost mythical, or at best traditional. For if the word 'penny' had been used to denote the pennyweight of the market, it would not have been described as 'round and without clipping.' It is pretty certain that the moneyers of Edward's time reckoned 24 grains to the penny. And the pennies of the reigns previous to Edward *do* weigh 24 Tower grains. And they weigh 25·6 Gothic grains, but not 32 of any grain that has ever been heard of.

We may affirm with certainty that Edward's pound sterling contained 5,760 Tower or Colonia grains, or 6,144 Gothic grains. It was exactly three-quarters of the Gothic pynd, and it was sub-divided into 20 × 12 × 24, whereas the old pynd was sub-divided into 16 × 16 × 32. But the Tower pound was the weight of pure silver worth 1*l*., and when issued as coin $\frac{1\frac{3}{8}}{18}$ fine it weighed 6,144 Tower grains, and this was the *trew pound*. That this was the trew pound mentioned by Edward I. is settled beyond doubt or cavil by the fact that he states distinctly that, though 20 shillings of money or drugs made a pound, 25 of everything else made a pound. Now, since there were 7,680 Tower grains in a pynd, we have only to take $\frac{20}{25}$—that is, $\frac{4}{5}$—of this to find the money trew pound. And $\frac{4}{5}$ of 7,680 is exactly 6,144.

Thus we see that in Edward I.'s day there were three recognised pounds—the old pynd, or market pound of 7,680 Tower grains (8,192 Gothic grains), the Tower pound of 5,760 Tower grains, and the trew pound of 6,144 Tower grains. Our present Troy pound did not exist till 1527, when it was established by Henry VIII. upon what must be described as a disgraceful arithmetical blunder, or as a piece of unworthy State dodgery and sharp practice deliberately planned with a view to extracting a penny in every pound for the Treasury. It weighed only 6,120 Tower grains.

There is no reason whatever to believe that the English people ever adopted the Roman sub-division of the popular or market pound (the Gothic pynd), which was always sub-divided into 16 ounces of 16 drams of 32 grains, and has so continued up to the present century, when, under the name of avoirdupois pound—a name it had long enjoyed—it was altered to 7,000 Troy grains. The Gothic grain had, it is true, long ago dropped out of use as too small a weight for most commercial transactions. But to contend, as some do, that there never was an avoirdupois grain is to ignore history and to trifle with common-sense.

The gold ligule does as a fact weigh *exactly* 5,760 Tower grains or a Colonia pound, and the standard silver cyathus weighs 6,144 Tower grains, or a trew pound. In using these Roman names instead of their Gothic equivalents, with which we are not certainly acquainted, it must be remembered that our ligule and cyathus are slightly larger than the Roman, though slightly smaller than the Greek μυστρον and κυαθος.

Let us call the old Teutonic pound the *pynd*. It was clearly used by the Saxons of Westphalia as well as by the Angles, Jutes, and Danes, and also by the North Germans or Prussians, long before the Saxon invasion of Britain. We have seen what it was—a cylinder 1 hand in diameter and half a hand in height, filled with standard wheat—that is to say, with wheat of a specific gravity of ·8. The pynd contained 8,192 grains; 1,024 such pynds make a pipe or yard cylinder. The yard cylinder was sub-divided on the binary scale without exception. The decimal and duodecimal scales were unknown. Thus an eighth of a pipe was a boll; an eighth of a boll was a peck; an eighth of a peck was a can; an eighth of a can was a gill. Probably all the halves and quarters had also names, though some of them dropped out of use and others took foreign names. Thus in Britain the double pipe became the *tun* and the double boll became the *cwm*, now spelt coomb, while the weight of the double peck became the *tod*. These are all three British names, and in all probability the old British measures which they denoted were slightly altered so as to fit them into the great Teutonic system. Similarly, when the Roman system with its mixed decimal and duodecimal scales came into conflict with the far finer binary system, the Roman measures were slightly altered, and the Roman measure names were conferred upon the old Teutonic measures. Thus it befell that the half-peck was called a gallon, or rather the late Roman gallon was altered to fit the half-peck, and the bushel became the half boll. The can (afterwards called a quart) contained 2 pynds, and the pynd was divided into two marks. Half a mark was a gill (by whatever name known); a quarter of a gill was an ounce. Hence an ounce contained 512 grains; and it was a cylinder half a nail in diameter and one nail in height. Take particular note of this measure. It is the one measure which the Romans were unable to alter even in Colonia. A sixteenth of an ounce was the famous penny, and the eighth of that was a styca.

OLD ENGLISH SILVER WEIGHT

Silver	Gothic grains	Mint grains	Troy grains
Pynd	8,192	7,680	7,228·23 a
Pound, Col.	6,144	5,760	5,421·177 b
,, Treu	6,553·6	6,144	5,782·578
,, Troy	6,528 c	6,102	5,760
,, Av.	7,933·3 d	7,437·5	7,000
Mark	4,096	3,840	3,614
Gill	2,048	1,920	1,807
Ora	512 ⎫	480	451·7648
Ounce, Col.	512 ⎭	480	451·7648
,, Troy	544	510	480
Shilling, Col.	307·2	288	271·0584
,, Treu	327·7	307·2	289·1289
,, Troy	326·4	306	288
Skilling	128	120	112·9412
Groat, Col.	102·4	96	90·3528
,, Treu	102·6	102·4	96·3703
,, Troy	108·8	102	96
Gothic penny	32	30	28·2353
Penny, Col.	25·6	24	22·5882
,, Treu	27·3		24·0041
,, Troy	27·2	25·5	24
Sceatta	16	15	14·1176
Styca	4	3·75	3·7
Grain, Goth.	1	$1\frac{5}{16}$	$1\frac{5}{17}$
,, Col.	$1\frac{1}{15}$	1	$1\frac{6}{17}$
,, Troy	$1\frac{2}{15}$	$1\frac{1}{16}$	.1

All these are the *weights* of the silver coins, and their *measures* are $\tfrac{1}{15}$th of the corresponding wheat measures. Thus a Tower pound of silver measures $\tfrac{1}{15} \times \tfrac{3}{4}$ pynd $= \tfrac{3}{5\,2}$ pynd. But a Tower pound of gold measures $\tfrac{1}{24} \times \tfrac{3}{4}$ pynd $= \tfrac{1}{32}$ pynd = half-nail cylinder, which *measure* of gold was the unit of Tower *weight*—the Colonia pound.

By actual measure, *a* 7,227·3 ; *b* 5,420·4797 ; *c* 6,529 ; *d* 7,984·5.

We have already seen what became of the Roman *millepassuum* in Britain. It has grown considerably in order to fit in with English land measures, and to serve as eight furlongs. Similarly, the *gallona*, or helmetful, was forced into the system as 8 pynds. Now, what would happen in Colonia when the Roman *libra* of 5,051 Troy grains (or, on Böckh's estimate, 5,053 grains) came into use as a money measure? It seems pretty certain that the Saxons would enlarge it in order to bring it into some simple relation with their own venerable pynd. Well, three-quarters of a pynd contains 5,421 Troy grains—a most important weight, as will appear. Hence the *libra* must swell out to fit this, just as the mile was drawn out into eight furlongs. Be it understood, we are now discussing the early history of the Tower pound or Colonia *pondus* and the avoirdupois pound or Teutonic pynd. It will be shown that this Colonia *pondus* was so adjusted as to become exactly three-quarters of a pynd, and so remained from the days of the Saxon invasion till the reign of Henry VIII.—not approximately but exactly, to within a grain. This will be seen when we come to the Act of Parliament which created the Troy pound in 1527.

In pre-Roman times the pynd, mark, mancus, ora, skilling, penny, and styca seem to have been used by Gothic peoples. But when the Saxons proper came over to England, they appear to have brought with them their Colonia *pondus* or money pound of three-quarters of a pynd. And they continued to sub-divide it in the Roman way instead of on the binary scale. Thus the *pondus* was divided into 12 *unciæ* or ounces (as the Troy pound still is). But since the pynd was divided into 16 ounces, it is obvious that the two ounces were the same weight. Hence the importance of the ounce. But whereas the Angles divided the ounce into 16 pennies, and the Romanised Saxons divided it into 20 pennies, the penny was reduced from 32 Gothic grains to 25·6 such grains. And the grain itself was enlarged so as to make it $\frac{1}{24}$ of the new penny. This is not the only time that the size of the grain has been altered. On both occasions it has been enlarged. On this particular occasion—namely, when the duodecimal scale with the Teutonic ounce for a unit was imposed on the Saxons by the Romans

—the ounce remained the same, while the pound was reduced to three-quarters of its original weight, and the penny was reduced to four-fifths of its original weight. The mark, which was 8 ounces, remained unaltered, 8 being a number which fits easily either into the binary or duodecimal scale. The *mancus*, perhaps, on the other hand, being the eighth part of a pynd, was reduced like the pynd to three-quarters of its original dimensions. Hence we hear of the *mancus* of 40 Saxon pence and of the *mancus* of 30 Saxon pennies, the latter being the money *mancus*. Ælfric distinctly says that 30 pennies make a *mancus*, on the same occasion in which he mentions the fact that 5 pence make a skilling. Hence he must have referred to the smaller or new penny of 24 Tower grains, or 25·6 old grains. And 30 times 25·6 is exactly 768 old grains, or an ounce and a half.

Returning to the penny, Athelstan in one of his laws says that 4 pence make a skilling. But Canute says 5, agreeing with Ælfric. Now a skilling was a quarter-ora, and contained 128 old grains. And that is precisely 4 × 32, and it is also exactly 5 × 25·6. Hence we are justified in supposing that the two kings were speaking of two distinct pennies—the old Gothic penny of 32 old grains and the new Saxon penny of 24 new grains or 25·6 old grains. Eventually the Saxon coinage with the Roman scale superseded the Gothic in all matters relating to the currency, but in other departments the Gothic system seems to have more than held its own till Henry VIII. was pleased or forced to legalise it under the singular name of haver-de-pois weight. The pound was still divided into 16 ounces, and the ounce was divided into 16 drams or Gothic pennies. And this brings us to the second change effected in the size of the grain.

Meantime, what was this Troy pound to which occasional reference is made, but which is not mentioned in the statutes till the time of Henry V.? Edward I. in one of his laws distinctly states that although 20 shillings make a pound of money and drugs, it takes 25 shillings of all other things to make a pound. Does this mean that a shilling (the weight) of flour was less than the shilling of silver?

Or that the pound of flour was heavier than the pound of silver? Of what pound can it be said that it contains 25 shillings? Let us see. Take the old pynd and divide it by 25. $\frac{8192}{25} = 327 \cdot 7$ old grains. But the money shilling of his day contained only 307·2 old grains or 288 new grains. Now, approximations are of no use; let us be exact to a grain.

We know that the Teutonic standard silver was $\frac{11}{12}$ fine. Therefore 307·2 grains of pure silver would when minted actually weigh 327·7 grains. Hence, when a shilling's worth of actual money was weighed it would be found to weigh exactly $\frac{1}{25}$ of the old popular pound or pynd, which was used in Edward I.'s time for all goods except money and drugs. And the real or trew weight of the actual money would naturally be called Trew or Troy weight by the people, to distinguish it from the theoretical weight used by the moneyers. And the people using their old Gothic weights would call a silver shilling the twenty-fifth of a pound (meaning the old popular pound of 8,192 old grains or 7,680 new or Tower grains). Then came Henry VIII. with his English proclivities and his contempt for the Roman religion and scales and measures, and his famous statute of 1527, in which Troy weight was first established. It is therein distinctly stated that the Tower pound of *fine* gold weighs only $11\frac{1}{4}$ *Troy* ounces. Now there was no such weight in either system as a Troy ounce. The popular ounce and the Tower ounce were one and the same. It is obvious, therefore, that the people used to call the twelfth of a Troy pound (or actual true weight of a pound of silver) a Troy ounce. But such a weight did not exist. If it had existed, it would have weighed an ounce and a fifteenth. Now the Tower or mint ounce (the real brass weight used in the Tower) was 480 mint grains. Hence the actual true or Troy weight of the coined money was 512 mint grains. But what did Henry do? He enacted that the pound Troy should exceed the Tower pound by three-quarters of an ounce. What ounce? The ounce existing both in the mint and in the markets, or the ounce which he himself was newly creating by enacting the Troy pound of 12 Troy ounces? Clearly the former. Thus, instead of adding 32 mint grains to the mint ounce, as he intended, he added only 30 mint grains, or $\frac{1}{16}$ of the only ounce then in existence. Consequently, instead of the trew pound referred to by Edward I., he invented a Troy pound 24 grains smaller.

This was the first time the ounce had been altered since the days when the great Teutonic system was perfected, long before the Anglo-Saxon invasion. And of course (the scale being still the same as before) the pound, ounce, shilling, penny, and grain were all altered in proportion. Let us follow the change in Tower or mint grains in order to avoid all the fractions necessary to express the weights in Troy grains.

The old Teutonic pynd weighed 7,680 mint grains. The Colonia or Saxon *pondus* weighed 5,760 mint grains, or just three-quarters of the former. The trew pound of 25 shillings mentioned by Edward I. weighed 6,144, and Henry VIII.'s Troy pound came to weigh (and still does) 6,120 mint grains, or just a mint pennyweight less than it should.

The actual words of the Statute of 1527 are these: 'And whereas heretofore the merchante paid for coynage of every Pounde Toure of fyne gold weighing 11¼ oz. Troye, 2*s*. 6*d*. Nowe it is determined by the King's Highness and his Councille that the foresaid Pounde Toure shall be no more used and occupied, but al maner of golde and sylver shall be wayed by the Pounde Troye, which maketh XII. oz. Troye, which exceedeth the Pounde Toure in weight 3 quarters of the oz.'

'Three-quarters of *the ounce*' is a somewhat ambiguous expression. But, whatever the intention, it is certain that three-quarters of an old mint ounce was added, and not three-quarters of the new ounce. But it is nevertheless an amazing fact that, in reversing the process in order to find the weight of the old Tower pound, all our writers subtract three-quarters of a *Troy* ounce from our *Troy* pound, and say that the Tower pound must have weighed 5,400 *Troy* grains, which it certainly did not. It weighed 5,760 mint grains, which is equal to 5,421·1776 Troy grains. This extraordinary error vitiates all the estimates of the weights of old coins, and utterly obscures the history of our weights and measures.

In order to verify this account of what actually took place, let us subject it to the most delicate and exacting tests.

THE FIVE POUNDS AND THEIR SUBDIVISIONS

	Goth. grains	Mint grains	Troy grains	
Grain, Goth.	1	$1\tfrac{5}{16}$	$1\tfrac{5}{17}$	Old grain
,, Col.	$1\tfrac{6}{18}$	$1\tfrac{6}{16}$	$1\tfrac{6}{17}$	Colonia or Tower gr.
,, Troy	$1\tfrac{7}{15}$	$1\tfrac{7}{16}$	$1\tfrac{7}{17}$	Hen. VIII.'s Troy gr.
Styca .	4 a	3·75	3·7	
Sceatta	16 a	15	14·1176	Probably!
Penny, Goth.	32 a	30	28·2353	
,, Col.	25·6	24 b	22·5882	Tower dwt.
,, Treu.	27·3	25·6	24·0941	Old Market dwt.
,, Troy	27·2	25·5	24 c	Present dwt.
Scilling, Goth.	128 a	120	112·9412	= 5 Col. pennies
,, Col.	102·4	96	90·3528	
Groat, Col.	102·4	96 b	—	Called English shilling
,, Treu.	109·2	102·4	96·3762	
,, Troy.	108·8	102	96	
Shilling, Col.	307·2	288	271·0584	} $\tfrac{1}{25}$ pynd Edw. I. trade shilling
,, Treu.	327·7	307·2	289·1289	
,, Troy	326·4	306	288	
Ora, Goth.	512 a	480	451·7048	} Never altered
,, Col.	512	480 b	—	
Ounce, Treu.	546·1	512	481·88	Edw. I.
,, Troy	544	510	480 c	Hen. VIII. till now
Mancus	7·8	720	677·64	} Gold coins of $\tfrac{56 \text{ tr. grs.}}{73 \text{ tr. grs.}}$
Byzant	1,024 a	960	903·5	
Gill	2,048	1,920	1,807	{ Ratio $\tfrac{\text{gold}}{\text{silver}}$ A.-S. = 12 : 1
Mark .	4,096 a	3,840 b	3,614	Not altered
Pound, Goth.	8,192 a	7,680	7,227·306307	Calculated 7,228·235205
,, Col.	6,144	5,760 b d	5,420·47973	,, 5,421·177
,, Treu.	6,553·6	6,144	5,782·578	
,, Troy	6,528	6,120	5,760 c	,, 5,759·259713
,, Av.	7,933·3	7,437·5	7,000	

When was this grain lost? When gold unit supplanted wheat unit, or when Troy pound came in? If we take the actual Troy pound of 7,227·306307 Troy grains in pynd measure, and take $\tfrac{12}{13}$ths of it, we get 7,679, and if we take the old pynd of 8,192, and take $\tfrac{15}{16}$ths of it, we get 7,680—the correct number. But this proves nothing, and perhaps the question is not worth answering. $\tfrac{1}{8000}$ is *small enough!*

a Gothic coins.   *b* William I. coins.   *c* Present Troy weights.
d Value, £1 Tower; weight, half-nail cylinder of gold.

A cubic inch of distilled water at 62° Fahr. (bar. 30°) weighs 252·458 Troy grains; and a yard cylinder contains ·785398 of a cubic yard; therefore the old pynd contains 35·7846964 cubic inches; therefore the pynd weighed 9034·1328837512 Troy grains (of water); and we know that a pynd of water = 10,240 old Gothic grains; therefore, the ratio between the old Gothic grain and the Troy grain is $\frac{9034}{10240}$ nearly; or, more accurately, ·88223954, call it ·88224. This is the ratio by actual measurement.

Now if my theory is correct the result should approximate very closely. Take away three-quarters of a mint ounce from our Troy pound. 5,760 − 338·8236 leaves 5,421·1704. And that is the Tower pound expressed in Troy grains; and this is three-quarters of 7228·2 Troy grains, which should equal the pynd; but the pynd contained 8,192 Gothic grains. Therefore, the ratio between Gothic and Troy grains should be $\frac{7228\cdot2}{8192}$ = ·88232, which is equal to the ratio obtained by direct measurement, within less than one grain.

State the case this way: since there are by actual measurement:—

9034·13288 Troy grains in a pynd of water,
and   7227·300307   ,,   ,,   ,,   std. wheat, sp. gr. ·8;
∴ there are 5420·47973   ,,   ,,   Tower pound,
and   5759·259713   ,,   ,,   Troy pound.
But there are 5760   ,,   ,,   ,,   ,,   as a fact;
difference   ·74   ,,   or less than three-quarters of a
                                   Troy grain in a Troy pound.

And since $\dfrac{\text{Gothic gr.}}{\text{Mint gr.}} = \dfrac{15}{16}$ and $\dfrac{\text{Mint gr.}}{\text{Troy gr.}} = \dfrac{16}{17}$ ∴ $\dfrac{\text{Gothic gr.}}{\text{Tr. gr.}} = \dfrac{15}{17}$

and $\dfrac{15}{17}$ 8,192 = 7,228·235295 = the number of Troy grains in a pynd of wheat at ·8 sp. gr., and 7,227·306307 is the *measured* number (see above). Difference, ·928988, or less than a Troy grain in the pynd of wheat.

It seems useless to question the theory, but it is simply amazing to find that the accuracy of the weights has been maintained for over fifteen centuries, in spite of the two great changes.

The Report of 1870 mentions a Roman pound of 5,759·2 grains, but it cites no authority. It says this pound was the $\frac{1}{125}$ of the large Alexandrian talent, but it does not say how and when this talent was measured to such a nicety. It will be observed that not only the number of grains but the first decimal figure are the same as mine, viz. 5,759·2, against 5,759·2597.

These two measures and the actual Troy pound of 5,760 are practically the same; that is to say, the difference between them can be accounted for by temperature, by barometric pressure, by the pureness of the water, or even by the accuracy of the measurer. What does it mean? Accident it cannot be. As for the Roman pound, it may be a guess and forced approximation—in fact, a lie. But figures cannot lie, and that the pound is based on the yard seems to me indisputable.

But *why* did the Roman Saxons of Cologne choose ¾ pynd? A pynd or half the hand cylinder is bad enough for a unit. Why not the hand cylinder or can itself? But three-quarters of a pynd or three-eighths of a can is so awkward for a unit that only the strongest reason could justify its selection. Now one-eighth of a can is the gill or nail-cylinder, and one-eighth of a gill is another cylinder—the half-nail. Is there any well-known substance of such specific gravity that a half-nail cylinder of it would weigh exactly a Colonia pound? There is, and only one that even approximates. And it is exact.

Specific gravity of gold is 19·2—that is, $\frac{water}{gold}$ is $\frac{1}{19\cdot 2}$

therefore a half-nail cylinder of gold weighs exactly a Colonia pound.

Thus a can of water weighs 18,068 Troy grains; a gill of water weighs 2,258·5 Troy grains; and a half-nail cylinder weighs 282·3 Troy grains. Multiplying by 19·2, which is the specific gravity of gold, we get 5,420·16 Troy grains. Therefore, a half-nail cylinder of gold weighs 5,420·16 Troy grains; and this is the weight of the Tower pound to within less than 1 grain. But if instead of 0,034

Troy grains to the pynd of water we take $9,034 \cdot 13288$, *the exact figure*, the half-nail cylinder of gold comes out $5,420 \cdot 47973$, which is *exactly* a Tower pound to the very last decimal figure. Can we doubt that the constructor of the Colonia pound took for his unit of money weight the half-nail cylinder of *gold*?—neither wheat nor water!

CHAPTER VII

BEFORE THE ACT OF 1824

ALTHOUGH, in our ignorance of the precise quality of wheat used by our forefathers as standard wheat, we are, at the outset, unable to compare modern and ancient weights with accuracy, we are under no such difficulty when we come to compare modern and ancient measures of capacity. We have only to assume that the yard has not materially varied in length, and we know to a grain what the old peck, the old bushel, and the old gallon must have contained. Thus a modern gallon contains 70,000 grains of water at 62° Fahr., and the old gallon, or cylinder of a quarter of a yard in diameter and an eighth of a yard high, must have contained 72,320 grains under the like conditions. Hence, we see that our modern imperial gallon is less than its prototype by some 2,320 grains, or approximately by one thirty-secondth of its bulk. The same proportion, of course, holds good of all the other binary measures of bulk. Thus, the nail-cylinder contains 8·946196875 cubic inches of water. Our gill contains 8·666 cubic inches, a difference of just over a quarter of a cubic inch. It seems a pity, when Parliament was altering the size of the gallon and substituting the imperial gallon for the various gallons in use seventy years ago, that it did not adopt as its standard gallon the half of the old quarter-yard cylinder. So far from creating any extra inconvenience, this arrangement would have been even more economical than the one adopted; for the new gallon would have been nearer to the old beer gallon than the imperial gallon is, viz. 286·7783 to 282, instead of 277·274 to 282.

Let us now see how far the customary weights and measures used in the different parts of England before the year 1824 justify the scheme set forth in this table. The yard cylinder doubled or double-pipe is called a tun, and it should weigh 2,048 pounds. According to the Report of 1820, a tun of wine is 2 pipes, equal to 252 gallons. According to the table it should be 256 gallons. The ton was mostly used as a weight of 20 hundredweights or 2,240 pounds. But we see that in Derbyshire it denoted 2,400 pounds, and in some agricultural districts as little as 1,709 pounds, and, in the words of the Report, 'in a ship's measure 40 cubic feet of timber are considered as a ton, being supposed to carry 2,000 pounds.'

A pipe is described as half a tun or two hogsheads or 126 gallons, and by an Act of Henry VI. it is said to be the same thing as a butt. A pipe of Jersey cider contained 120 gallons. According to the table it ought to contain 128. By an Act of Queen Anne the butt of wine or cider is said to contain 2 hogsheads. A cask of Gloucestershire cider contained 110 gallons. It seems probable that the pipe, butt, and cask were different names for the same thing—namely, 128 gallons. The vat or fat of corn was 8 bushels, which is half a pipe; but by an Act of George III. a vat of coals was made to contain nine bushels. Again, the quarter was universally regarded as containing 8 bushels and as being the quarter of a tun. But even this measure varied according to the locality and to the substance measured. For instance, we are told that a quarter of chopped bark contained nine heaped bushels in some parts of Yorkshire; and in Banff it denoted 8 bushels and 3 pecks. There can be little doubt that a hogshead was the same measure of 64 gallons; but, as we have seen, an Act of Henry VI. declared it to consist of 63 gallons, and in Devonshire a hogshead of lime contained 72 gallons. Oddly enough, although a hogshead of Jersey cider contains 60 gallons, a hogshead of Herefordshire cider contained no less than 110 gallons. And at one time or another the hogshead of ale or beer has denoted 48 or 54 gallons. This same measure of 8 bushels is known in Wales as a peget. The Report of 1820 gives coom or coomb as half a quarter, equal to 4 bushels, and the cran of herrings is declared by 41 Geo. III. to contain 34 wine gallons.

The word 'barrel' has varied in meaning considerably; but there is little doubt that it originally stood for the same measure as the coomb—namely, 32 gallons. It is declared by 43 Geo. III. to mean 36 gallons of ale or beer, but in London before this Act the barrel of ale was only 32 gallons, of beer 36; and in the country both were 34. According to another Act, 38 Geo. III., a barrel of beef was 32 wine gallons. By 5 Geo. I., a barrel of cod-fish was 32 gallons, but by 18 Geo. III. a barrel of coals was declared to contain not 4, but 3, Winchester bushels. A barrel of honey was made to contain 32 wine gallons by 23 Eliz. Charles II. made a barrel of nuts 3 bushels instead of 4. A barrel of spirits was allowed to contain $31\frac{1}{2}$ wine gallons, possibly on account of its liability to evaporation. A barrel of vinegar is declared by William III. to contain 34 ale gallons; 32 gallons was understood by a barrel of beef or of herrings in Scotland; and in Ireland a barrel of grain meant 4 bushels, and a barrel of potatoes was taken as a weight of 20 stone or 280 pounds, whereas a barrel of standard wheat weighed 256 pounds. On the whole survey, there can be little doubt that the barrel, like the coomb, contained 4 bushels or 32 gallons.

The pack seems originally to have denoted a weight of 256 pounds—that is to say, the weight of a coomb of standard wheat. It seems to have fluctuated a good deal. A pack of yarn was 480 pounds in some parts, but in Huntingdonshire a pack of wool was 240 pounds and a pack of flax in Kent was also 240 pounds. I think when we observe that one of these packs is exactly the double of the other, we shall be disposed to accept the explanation suggested on p. 59. The Welsh hobed seems to have been a measure of 4 bushels. We now come to the sack; this also was originally a measure of 4 bushels and a weight of 256 pounds. By an Act passed in the third year of the reign of George II., the sack itself is ordained to measure 50 by 26 inches. This would, if closed at top, contain a little over 3 bushels. Here again, as in the case of the pack, there would seem to have been two sacks, one the double of the other; for a sack of wool was declared by 14 Ed. III. to weigh 2 weys or 26 stone. In the reign of George III. the length of the sack was increased to 52 inches. As might have been expected, when used as a weight measure the sack varied considerably, and what is more remarkable is that it varies so much as a measure of capacity also. In Essex a sack of charcoal was 8 pecks, in Gainsborough a sack of corn was 2 bushels, and in Bedfordshire 5 bushels. In Gloucestershire a sack of potatoes was 3 bushels, and in Worcestershire a sack of apples was 4 bushels. On the whole, the sack seems to have denoted a measure of 2 bushels or of 4. In Devonshire a bag of wheat meant two bushels weighing 140 pounds. But in Shropshire it contained 3 customary bushels, and in other parts of the country both more and less. In the north country the boll was used to signify a measure of two bushels, and it is quite possible that this is the old English name for the strike, or eighth of a pipe. But in Scotland the term is used to denote a measure of 4 firlots or nearly 6 Winchester bushels, and that may account for the fact that in some parts of Northumberland the boll was actually used as a measure of 6 bushels for barley and oats, and the same measure was used in the Isle of Man. The name of the liquid measure of the same capacity is, as we know, the kilderkin, whose history we have already traced. The strike is another of our measures which varied in its contents from half a bushel to 1, 2, and 4 bushels, though it eventually seems to have given its name to the measure of 2 bushels.

This brings us to the bushel. At the beginning of this century there seem to have been over a dozen lawful bushels of wheat. The most usual was the Winchester of 8 gallons or 32 quarts, but in Newcastle it contained 36 quarts, in Burton-on-Trent 35 quarts and a pint, in Whitchurch 38 quarts, in Hereford, Montgomery, and many other places 40 quarts, in some places 33 quarts, in Leeds 34, in Fishguard 43 quarts, and in many places it was used as a measure of weight. In some parts of Yorkshire it denoted a weight of 70 pounds; in Chester 75 pounds; in North Wales 84 pounds. In short, if you asked for a bushel of wheat at the beginning of this century you might have got 64 quarts at Tavistock or 34 quarts at Barnstaple, while in Cornwall a bushel contained no less than 3 Winchester bushels. In the main, however, it seems clear that the original bushel contained 8 gallons, and weighed 64 pounds in standard wheat; the same measure for liquids was called by its Dutch name of firkin.

Half the weight of a bushel of standard wheat—namely, 32 pounds—was called a tod, and when the stone was reduced from 16 to 14 pounds the tod became 28 pounds, and the wey of 128 pounds came down at the same time to 112. All over the country in 1820 the tod seems to have been 28 pounds, but in Sussex and the Channel Islands it still remained 32 pounds, and in some parts of Yorkshire it was reckoned at $28\frac{1}{2}$. Another name for this weight was the quarter-wey, which was afterwards shortened to quarter. The Welsh name for the half bushel or double peck was kibin.

The peck was the span cylinder, or cylinder of 9 inches diameter and 9 inches in height; it contained exactly two gallons or a quarter of a bushel, and its weight in standard wheat was 16 pounds, which was called a stone. A cylinder of the same diameter and half the height formed the gallon, and the weight of a gallon of standard wheat was called a clove. When the stone was reduced to 14 pounds, the clove or cloven stone was of course reduced to 7. The old true gallon contained 286·7 cubic inches, but the wine gallon in the time of Queen Anne contained 231 cubic inches, and was described as a cylinder 7 inches in diameter and 6 inches high, which shows that its origin and its relation to the yard had been entirely lost sight of. By 45 Geo. III., the Winchester gallon is estimated at $272\frac{1}{4}$ cubic inches. At the custom house a gallon of honey was reckoned as 12 pounds; a gallon of train oil at $7\frac{1}{2}$ pounds; the Guernsey gallon measured 252 cubic inches; and our modern imperial gallon measures 279·272 cubic inches, or about $7\frac{1}{2}$ cubic inches less than it should do. According to the Report of 1820, the pottle does not appear to have meant more or less than half a gallon in any part of the country, and the same measure was frequently known by its Dutch name of stoup. Although there can be no doubt of the former existence of two stones, just as we now have the butcher's stone of 8 pounds and the ordinary stone of 14, it may be as well to see what other stones were in use in the early part of this century. A stone of lead weighed 12 pounds; of wool, 14; of hay, tallow, and yarn in Cumberland, 16; and of wool in Yorkshire the same weight, and of butter in Westmorland; in Herefordshire a stone was 12 pounds, and in Liverpool 20 pounds. The commissioners found provincial weights called stones in various markets weighing 4, 5, 6, 7, 11, 13, 14, 15, 17, 18, 21, 22, 24, and 26 pounds. However, in spite of all these variations, there is not a shadow of doubt that the original stone was 16 pounds and the clove stone 8 pounds.

The quart has always been understood to mean a quarter of a gallon or 2 pints. The old English name for this measure was probably can; the term 'quart,' however, has been used as a weight of 3 pounds, and in the Isle of Man as a weight of 7 pounds, of wool. The can seems to have been so far superseded by the quart that the commissioners of 1820 do not even mention it, though they mention the canter as the name of a quart of ale in Bedfordshire; they also mention the tankard as an ale measure of 1 quart. The pint was uniformly understood to be half a quart, and it does not appear to have been used in England in any other meaning; but in Bedfordshire a pint was usually called a mug. The weight of a pint of standard wheat was originally taken as the standard of weight and called by the same name, pynd; but, as we have seen, the pynd differentiated into the pint measure and the pound weight.

Below the pint our measures of capacity are not numerous: we have the gill, which is the quarter of a pint, and which originally was a cylinder measuring a nail in diameter and a nail in height. And a gill of standard wheat weighed 2048 old grains. The commissioners do not mention the nipkin or the pipperkin, but they define the drachm as the hundred and twenty-eighth part of a wine pint or eighth of a fluid ounce, and the drop as a measure they define as the sixtieth part of a drachm when dealing with water, but as half that measure when dealing with tinctures.

Below the pound weight we have of course the ounce, pennyweight, and grain, and also the dram, besides the apothecary's measures; but when we reach these small weights we become entangled among the earlier and later coin-weights, and it is difficult to say with any degree of precision what these old weights were in terms of those of the present day. One may of course conjecture that the carat or κεράτιον (as some derive it) was the quatre or four grain weight; that the dram and the thrimsa had something in common: but all this is little better than conjecture.

Not content with establishing general weights and bulk measures, the State has made special regulations and established special measures for various commodities. Most of these were repealed by the Act of 1824.

Amongst the general commodities selected by the State for careful attention are bread, ale, wool, wine, honey, oil, corn and malt, cheese, malmsey, fuel, salt, glass and earthenware, sweets, vinegar, fruit, coals, timber (beechwood in particular), wax, eels, herring and salmon, barrelled fish (red and white herrings in particular), linen and flax, hemp, lime, cloth, gold and silver, beef, mutton and veal.

One would have thought that a gallon of honey would be the same size as a gallon of oil or wine, and that the general measure would serve for all; that a yard of cloth would be as long as a yard of rope, and no longer; that a wey of cheese would weigh the same as a wey of wool. Not at all; such was the confusion brought about by a system which confounded weight with bulk, that frauds of every description had to be provided against. Dry cloth well stretched will shrink 2 yards and more in the piece, and the State, therefore, enacted that all cloth should be watered 4 or 6 hours before it was sold, and not afterwards stretched; and it visited with heavy penalties all those whose cloth shrank more than 1 yard in a piece. Again, small fishes like herrings in a 32 gallon barrel fit closer than large fishes like salmon; of this the State took cognisance, and varied the size of the barrel.

In measures of length, the custom of interposing the thumb had been so universal that the thumb came to be considered as part of the measure; and in process of time an inch was substituted for it, so that the yard was made to consist of 37 inches. And by an Act of Queen Anne, each yard is to have an inch added to it, 'instead of that which is commonly called a thumb's length,' and, shortly after, a table 30 yards long was ordered to be provided at Blackwell Hall, each yard counting 37 inches.

CHAPTER VIII

GREEK MEASURES

LET us now turn to the Greek measures. Consider what we are asked to believe: first, that there were three systems of weights and measures existing side by side for centuries, with the same names, the same sub-divisions, and wonderfully related; secondly, that the chief weight measure is based on a foreign bulk measure—the Roman cubic foot; thirdly, that the chief bulk measure is out of all harmony with the rest of the series—a peculiarly simple one—and that it also is based on another foreign measure, the Babylonian foot; fourthly, that the following absurdly improbable ratios subsist between the Æginetan, the Euboic and the Roman bulk measures: 2, 3, 4, 5, 6. We are actually called upon to believe that the Roman *amphora* was to the Euboic ἀμφορεύς as 2 to 3; to the Euboic μεδιμνος as 2 to 4; to the Æginetan ἀμφορεύς as 2 to 5; and to the Æginetan μεδιμνος as 2 to 6. In all this the hand of the German theorist is but too visible. Perhaps as a sufficient confutation of such a grotesque theory it may be pointed out that the common measure of these five is a 3-gallon jar, which measure is itself nameless in every one of the systems, whereas the 2-gallon jar is common to *all* the old systems, as the ἑκτεύς, *modius*, or peck. How the blunder arose seems to me pretty obvious, although it has hitherto escaped detection. The old field measure which we call the peck, and which is a span cylinder, was well known to both Greeks and Romans, but, as we know, the Greek span was a little longer than the Roman, viz. as 25 to 24. This difference would be imperceptible to the most experienced eye.

Hence the ἑκτευς and the *modius* were commonly regarded as the same measure, whereas as a fact the *modius* was only eight-ninths of the ἑκτευς. When the ἀμφορευς came to be measured by the Roman *modius* it was found to hold $4\frac{1}{2}$. And our Roman informants jumped promptly to the conclusion that it contained $4\frac{1}{2}$ ἑκτεις,—a most inconvenient fraction, and altogether out of keeping with the simple Greek binary system. Obviously the ἀμφορευς was 4 ἑκτεις exactly. Thus the ratio of the *amphora* to the ἀμφορευς (2 to 3) came about in this wise, and it is in no way affected by the correction. The *amphora quadrantal* contained practically 3 *modii* (a *very* small fraction more), and the ἀμφορευς contained 4 ἑκτεις—*i.e.* $4\frac{1}{2}$ *modii*. Hence the former was two-thirds of the latter. The same correction must be applied to the ξεστης, which was one-eighth larger than the *sextarius*, though usually supposed to be the same measure.

What was the μεδιμνος? I do not know of any ancient Greek writer who compares the ἀμφορευς and the μεδιμνος. The later Romanised Greeks are, as we have seen, not to be relied on Why, after doubling fourteen times and keeping up the binary scale from one end of the scale to the other—from the κοχλιαριον or eggspoonful to the ἀμφορευς or bushel—the Greeks should suddenly adopt a measure of $1\frac{1}{2}$ ἀμφορεις and call it a μεδιμνος I fail to see. Surely the two words are either two names for the same measure, or one is double of the other. We ourselves discriminate between our corn measures and our liquid measures by giving them different names, although they are, or originally were, the same. We speak of a bushel and a firkin; of a quarter and a hogshead. Why should they not speak of a μεδιμνος σιδηρος and an ἀμφορευς μετρητης?

But various contemporary Roman authors tell us that the Attic μεδιμνος contained exactly 6 Roman *modii*. This would be $5\frac{1}{3}$ ἑκτεις. And we are told that the Attic μεδιμνος was two-thirds of the Æginetan. Now this latter is the measure we wish to get at. It belongs to the old Greek pipe series. There may have been a dozen others of the same name, just as there were over a dozen bushels in England at the beginning of this century, some twice as big as others. But we do not particularly want to know much of these degraded or corrupted measures. Well, three-halves of $5\frac{1}{3}$ ἕκτεις is the Æginetan μεδιμνος. And three-halves of $5\frac{1}{3}$ is exactly 8. Here we have it. The original Greek μεδιμνος contained 8 ἕκτεις. In other words, it was a double-bushel; call it a strike, or boll, or kilderkin. Whatever we call it, it was a πηχυς cylinder. Why the Attic μεδιμνος should have been reduced to two-thirds of this I cannot tell. The Sicilian is said to have been one-sixth less still. The explanation of these divergences may safely be left to antiquarians. We are merely dealing with metrology. It is also likely enough that writers familiar with the new Roman measures would read that a μεδιμνος contained 2 ἀμφορεις, and would jump to the conclusion that it contained 2 *amphoræ* or 6 *modii*. If this is the correct explanation of the statement (as seems probable), then the Attic and ordinary Greek or Æginetan μεδιμνος were one and the same.

We must not forget that the later Greco-Roman physicians, to whom we are indebted for much of our knowledge of the ancient Greek measures, wrote in Greek and thought in Latin.

For example, when they wrote ἀμφορευς they meant *amphora*; when they wrote χους, they frequently meant *congius*.

This is the simple explanation of the fact that modern metrologists attribute to the Greeks the adoption of an unwieldy and anomalous bushel. The *amphora quadrantal* was three-fourths of the old *amphora* or Roman bushel, which was eight-ninths of the Greek ἀμφορευς. Hence the quadrantal was two-thirds of the ἀμφορευς. Similarly, the *congius*, being the eighth of a *quadrantal*, was the twelfth of an ἀμφορευς. But the true χους of the early Greeks fell naturally into the old binary series. Again, the κοτυλη is one of the measures of the old pipe scale; and that form of it which is described by the later writers as the twelfth of a χους is the Roman *cotyla*, which is the twelfth of the new *congius*. This probable explanation is strengthened by the testimony of Suidas, who says there were two χοες, which he distinguishes as the χους and the χοευς, one of which contained (he says) 6 ξεσται and the other only 2. The proportions of the smaller χους are, as might be expected, wrong. A χους of 2 ξεσται would, of course, be a χοινιξ. The real old Greek χους contained 4 ξεσται, whereas the new *congius* contained 6 *sextarii*. The χους in the ordinary tables is obviously a foreign measure.

It is simply incredible that the early Greeks could have used a field measure containing exactly two Roman *amphoræ*, and bearing no simple relation to their own ἀμφορευς. And yet several respectable Roman authorities are agreed that the Attic μεδιμνος contained exactly six *modii*, which equals 5½ Greek pecks. Is it likely that such a measure ever existed? I confess this well-attested proportion is difficult to dispute, and more difficult to explain. But of this we may be sure, the Greek μεδιμνος was either one bushel or two—that is to say, one ἀμφορευς or two. I am strongly of opinion that the αμφορευς and μεδιμνος were one and the same measure, like our bushel and firkin; the one name used in connection with grain and dry goods, and the other in connection with liquids. This view is again supported by the authority of Suidas, who says the μεδιμνος contained 108 λιτραι. Commenting on this statement, Mr. Philip Smith says: 'confounding it apparently with the μετρητης (ἀμφορευς), the chief Greek fluid measure.' Yes, and why not? Surely a Greek writing eight centuries nearer the time would be less likely to make such a confusion without just cause than modern foreign critics almost wholly dependent on slipshod physicians writing after the conquest of Greece. Besides, this number 108 is remarkable; for if 108 λιτραι make an ἀμφορευς, clearly 96 would make an old *amphora*, and 72 a new *quadrantal*, which is 48 *sextarii*. Hence the λιτρα comes out as exactly three-fourths of a *sextarius*, just as the Colonia *libra* was three-fourths of a pynd. This seems to confirm my hypothesis.

We are now in a position to say exactly to a grain what was the weight of the Æginetan talent. It was the weight of a Greek bushel or ἀμφορευς *of water*, and the bushel was a cylinder a πηχυς in diameter and ½ πηχυς high, and the weight of this bulk of water is 597,840 Troy grains. The μνα or sixtieth part of this is 9,964 grains, and the δραχμη 99·6, a figure which is nearer by some 12 grains to the weight of the Æginetan δραχμαι actually found than Böckh's estimate of 112 grains. They average about 96 grains.

The Euboic talent was 498,200, being the precise weight of an ἀμφορευς of wheat at 66·6 pounds to the English bushel or $\frac{10}{9}$ths. The sixtieth or μνα would be 8,303·3 Troy grains, and the Euboic δραχμη would be 83 grains. No reliance whatever can be placed on any of the coins found and believed to be Euboic.

The Solonian talent was 358,704, being the weight of a χοινιξ of pure gold, of which the specific gravity is 19·2. In plain English, it is the weight of a Greek quart pot of gold, which is about one-thirtieth larger than an English quart pot or kan. The μνα was 5,978·4, and the δραχμη 59·7. This is the greatest difficulty, for we are told that the later Attic δραχμαι actually found average about 66½ Troy grains; but as I have said, these coins may not be what they are supposed to be.

All these measures accord exactly with the testimony of the Greek writers, according to whom the ratio of the three talents was

$$\left.\begin{array}{l}\text{Æg.}\\\text{Eub.}\\\text{Solon.}\end{array}\right\} = \left\{\begin{array}{l}120\\\overline{100}\\\overline{72}\end{array}\right.$$

The Solonian money is said to have been about $\frac{88}{89}$ fine. Now we know the difficulty of extracting the whole of the alloy out of silver. Hence we must not expect to find that the Greeks knew the precise specific gravity of pure silver.

A Greek κυαθος of pure silver would weigh 6,087 Troy grains, and the Solonian μνα contained 5,978 Troy grains; difference, 109 grains. This is not exactly a sixtieth, but it is about a fifty-sixth, and, having regard to the uncertain specific gravity of the silver, we are justified in suggesting that the Attic silver μνα was the pure silver contained in a κυαθος of standard silver $\frac{88}{89}$ fine.

A small cylinder one nail in height and half a nail in diameter— in other words, a κυαθος of standard silver—would weigh and be worth exactly one μνα. Of course, when I say 'nail' I mean the Greek nail or half-σπιθαμη. These cylinders would, it is hardly necessary to point out, be much more convenient, both as weights and perhaps as money, than the old-fashioned method of weighing or measuring the silver for each separate transaction. Not only do we know that this *mina* was called the *silver mina*, but we know that it became quickly popular, which is unusual with new measures, unless they possess some conspicuous advantage over those in use. Above all, as we shall see, the Roman *cyathus* of standard silver at what we know to have been the Colonia fineness, namely $\frac{16}{18}$, contains exactly 5,051 Troy grains, the precise weight of the Roman mint *libra*, which seems to have been about 62 grains heavier than the market weight. Thus by the adoption of his new talent, Solon achieved three very important objects: he got a talent of one quart of gold, he got a *mina* equal to the hundredth of a bushel of water exactly, and he got a κυαθος of standard silver containing exactly the new *mina* of pure silver. There could hardly be any other reason for making such a revolutionary change as is involved in reducing the money talent to three-fifths of its original weight.

People were thus enabled to pay their debts by simple measure in the ordinary bulk measures of the country. Thus if anyone owed 16 *minas*, he would measure out a ξεστης of standard silver. I suggest that it was this which led to the new system being known as the Attic Silver System. Fifteen gills (τεταρτα, *quartarii*) of standard silver would contain exactly a talent of pure silver.

The further we look into these matters the more we are compelled to dispute Mr. Grote's statement that the Greek bulk measures and the Greek weights have no definite relation to each other. For not only is the Æginetan talent simply the weight of a Greek bushel of water, but the Euboic talent is the weight of a bushel of wheat. Then the Solonian talent is the weight of a kan (χοινιξ) of gold; and now it appears that the Attic silver *mina* was nothing else than a κυαθος of silver. All these simple relations have hitherto remained hidden from view by reason of the careless blunders of writers living after the Roman Conquest, especially the confusion of the ἑκτευς and the *modius*.

There can be little doubt that a people with a well-developed binary scale of bulk measures would possess a similar scale of weight measures, sub-divided in the same way. This was the case with the Goths, as we have seen. Assuming that the *mina* was originally $\frac{1}{64}$ of the Æginetan talent, it would be the weight of a ξεστης of water. We are told that there were 60 or 50 manch in a kikkar (both in the Jewish and Babylonian systems), 'according to the mode of computation.' What that mode of computation was we are not told. We shall have little difficulty in perceiving that the bushel contained 60 pounds of water and 50 pounds of wheat, and that consequently the specific gravity of standard wheat in the East was $\frac{5}{6}$. Wherever the water scale prevailed in regard to weights, the *mina* would be the weight of a ξεστης of water ; and in those places where the wheat scale prevailed (agricultural districts) the *mina* would be the weight of a ξεστης of wheat. In the one case we have the Æginetan weight, and in the other the Euboic. By this simple arrangement the old Babylonian weights would fall in with the Greek pipe scale.

But for purposes of counting as distinguished from measuring, binary numbers are less convenient figures than round numbers. Now $64 \times 96 = 6,144$, while $60 \times 100 = 6,000$. Hence by slightly altering both weights we obtain the simple Oriental scale for counting. And the δραχμη is increased by about $\frac{1}{40}$. This is only about 6*d.* in the £—not a great price to pay for a simple system of weights.

Concerning the smaller Greek and Roman bulk measures we are in a position of some doubt and perplexity. We are indebted for nearly all we know, or think we know, about them to the Romanised Greek physicians of the first two centuries of our era. And these writers are all influenced by the Roman duodecimal scale which prevailed in their day, and which, as I have said, was not used by the Greeks before the Roman invasion ; although some of the commoner Roman measures were as familiar to the Greeks as the French measures are to us. Just as we speak of a French quart bottle, meaning a litre, so they spoke of a Roman χους, meaning a *congius*. It by no means follows that the measures were the same, or even very nearly equal.

From the ἐκτεύς or peck down to the ὀξύβαφον or half-gill, we are on firm ground. Here the binary scale reigns supreme and unquestioned (except in the case of the χοῦς). But after the ὀξύβαφον, *acetabulum, vinegar-cruet*—or, in plain English, wineglass—we come to the κύαθος. At this point we are suddenly told that this measure is the twelfth part of the ξέστης, a statement which is utterly incredible. The explanation is simple. The Romans, like every other people, chose the *sextarius* as their *as* or chief unit, and they divided it into twelfths; and the *cyathus* was selected to do duty for the *uncia*, being thus promoted from a sixteenth to a twelfth. However, the authorities are forced to admit that even in Rome 'the use of the *cyathus* as the *uncia* of the *sextarius* appears to have originated with the physicians in later times.' This seems to me to be going too far. But it is correct to say that the use of the κύαθος by the later Greeks as the twelfth of the ξέστης did so originate.

The change took place in Rome probably at the time of the adoption of the *amphora quadrantal*, which seems to have occurred at the date of the Sillian plebiscitum, whatever that date may be. On that occasion the *sextarius* underwent no change, or hardly any. The *amphora* was reduced from a bushel (a cylinder half a cubit in height and half a cubit in radius) to a cubic foot. The *congius* was taken as a 6-inch cube. The *sextarius* remained 36 cubic inches; the new *cyathus* was promoted to 3 cubic inches, and in all probability the official *ligula*, which was, I think, formerly not a quarter- but a half-*cyathus*, became the third of the new *cyathus*—namely, 1 cubic inch. But this change was never effected in Greece until long after the Roman Conquest; in fact, not for a century or so after Christ. We may therefore safely place the κύαθος in the table as half an ὀξύβαφον.

Our difficulties now begin. There are the names of four smaller measures to place:—the κόγχη, the μύστρον, the χήμη, and the κοχλιάριον. Smith's *Dictionary of Antiquities* makes 10 κοχλιάρια go to a κύαθος, but Rhemnius Fannius (or else Priscian) says 24 made a

H

κυαθος. Bearing in mind that these two writers lived after the Romanisation of the Greek κυαθος, we must understand them to say that 24 κοχλιαρια made a *new* κυαθος (the twelfth of a ξεστης). Therefore we shall not be wrong in assuming that it was the $\frac{1}{12}$ of the old κυαθος. In other words, this little spoon or limpet-ful was the $\frac{1}{288}$ part of a ξεστης. In the pipe scale this measure is the dram.

Referring to Hussey's *Ancient Weights* we read that there was a large χημη and also a small one, bearing the relation of three to two. Of course there would be as many sizes as there were χημαι at first, just as there are countless sizes of jugs and kettles now; but it is rather provoking to be told of *every* one of the smaller Greek measures that, after becoming technical and definite, there were two sizes. No business could have been conducted with such an absurd and unnatural system of measures. The small χημη, we are told, held two κοχλιαρια, which is likely enough. We have now to deal with the κογχη and the μυστρον, and here we are met on the threshold with the usual story. Galen says there was a large and a small μυστρον; but as he proceeds to contradict himself several times on the subject, we may let him pass. Cleopatra—not the queen, but the lady who knew all about cosmetics—writing about the first century of our era, says there were two μυστρα, the big one being $\frac{1}{16}$ of a κοτυλη, and the little one $\frac{1}{24}$. This last fraction throws an unexpected light on the whole matter, and the unusualness of the fraction lends a colour of truth to the statement. The $\frac{1}{16}$ of a κοτυλη is exactly half a κυαθος. Now, what fraction would this be of a κοτυλη made to contain 8 *new* κυαθοι? Just $\frac{1}{21\frac{1}{3}}$, or practically $\frac{1}{24}$. Thus Cleopatra would learn that a μυστρον might be either $\frac{1}{16}$ or $\frac{1}{24}$ of a κοτυλη.

There is still a vacant place in the binary table, and there is still a well-known measure seeking a place. It might be pardonable to stick the said measure in the said place without more ado. But let us inquire. Κογχη is a liquid measure 'of which there were two sizes.' Of course. We are quite prepared to hear that. But on this occasion the sizes are so different as to be appalling. The smaller, we are told, was half a κυαθος and the other was three times as large, if not four. Whichever of these measures we accept, there is clearly no room for the κογχη. If we believe that the κογχη was half an old κυαθος, then we must either degrade the μυστρον to a quarter or assume that they were different names for the same measure, which is quite possible. But, since Galen says the large μυστρον was $\frac{1}{18}$ of a κοτυλη, thereby nearly agreeing with Cleopatra, who says $\frac{1}{16}$, I think we must admit the claim of the μυστρον to be a half-κυαθος. And the κογχη, being so variable or indefinite as to represent either an ὀξυβαφον or a third of that measure, may be taken to be the quarter of a κυαθος, in so far as it was ever at any time a definite measure at all.

This gives us the following complete Greek binary table:—

Greek Pipe Scale

Greek measures	Κοχλιαρια	Ξεσται	English equiv.
Κοχλιαριον	1		Dram
Χημη	2		Fl. drachm
Κογχη	4		Skilling
Μυστρον	8		Hlf.-nail Cylr.
Κυαθος	16		Fl. ounce
Ὀξυβαφον	32		Wineglass
Τεταρτον	64		Gill
Κοτυλη	128		Mark
Ξεστης	256	1	Pynd
Χοινιξ	512	2	Kan
Χους	1,024	4	Pottle
Ἡμιεκτον	2,048	8	Gallon
Ἑκτευς	4,096	16	Peck
—	8,192	32	Tod
Ἀμφορευς	16,384	64	Bushel

The κοχλιαριον, if equal to the Roman *cochlear*, was an eggspoon. 'Sum cochleis habilis, nec sum minus utilis ovis' (Martial, xiv. 121). We are not compelled to believe that this is how it got its name.

We should thus be bound to assign similar explanations to the κογχη and χημη. When technically employed as a fluid measure for drugs it would be the sixteenth part of a κυαθος, or half our fluid drachm. And the weight of this measure of standard Gothic (or Roman) wheat is a dram in the pipe scale, and that weight of silver is a Gothic penny.. Two κοχλιαρια made a χημη or fluid drachm. Four made a κογχη, and 8 made a μυστρον, while 16 made a κυαθος, as Cleopatra says. Now the μυστρον and the Roman *ligula* (or *lingula*) were probably the same thing—a kind of split bone, or what we call a lobster-spoon. It was like a little tongue, and it was used for getting the marrow out of marrow-bones. According to some of the later writers, it was said to be a quarter of a *cyathus*, but we have seen that at that time it was probably a third. No people with a simple duodecimal system of measures would divide three cubic inches into four parts. The only question therefore is, What was the *ligula* before the new *congius* was invented? We have seen that there is good reason to believe that the μυστρον was half a κυαθος, and that the *ligula* and μυστρον were the same utensil. But what is more to the purpose is the fact that the old *ligula* and the new official *ligula* (the cubic inch) would be very nearly equal, provided that the old one was half an old *cyathus*. That is to say, the *quartarius* not having appreciably varied, it was formerly divided into two *acetabula*, four *cyathi*, and eight *ligulæ*. It was afterwards divided into three *cyathi* and nine *ligulæ*. Thus the new *ligula* was eight-ninths of the old—a not very serious or inconvenient change.

The Greek Grain —When Solon established his gold talent of six-tenths of the Æginetan, of which there is no doubt whatever, it was divided into 60 *minas* and each *mina* into 100 drachms. The drachm was divided into 6 obols, and if the Solonian obol was divided into any still smaller parts we should suppose it would be divided into *ten* somethings. Call them grains. This would make the drachm consist of 60 Greek grains. Now Solon's new money is said to have been $\frac{59}{60}$ fine. That is, it contained exactly one such grain of alloy in a drachm. Some of the Solonian coins have been tested, and found to contain precisely

GREEK BULK MEASURES AND WEIGHTS

										Greek gr. wheat at ·8 a	Greek grains of water[a]
Ὀξύβαφον	.	.	.	.	.	.	.	.	.	937·5	1,171·875
2	Τέταρτον	.	.	.	.	.	.	.	.	1,875	2,348·75
4	2	Κοτύλη	.	.	.	.	.	.	.	3,750	4,687·5
8	4	2	Ξέστης	.	.	.	.	.	.	7,500	9,375
16	8	4	2	Χοῦνιξ	.	.	.	.	.	15,000	18,750
32	16	8	4	2	Χοῦς	.	.	.	.	30,000	37,500
64	32	16	8	4	2	Ἡμίεκτον	.	.	.	60,000	75,000
128	64	32	16	8	4	2	Ἑκτεύς	.	.	120,000	150,000
512	256	128	64	32	16	8	4	Ἀμφορεύς or Μέδιμνος	.	480,000	600,000

$$\text{Æginetan talent} = \text{Ἀμφορεύς of } water = 600{,}000 \quad \text{Greek grains} \qquad M\nu a \begin{array}{c}\text{grs.}\\10{,}000\end{array} \quad \Delta\rho. \begin{array}{c}\text{grs.}\\100\end{array} \quad O\beta. \begin{array}{c}\text{grs.}\\16\tfrac{2}{3}\end{array}$$
$$\text{Euboic} \quad \text{,,} \quad = \text{Ἀμφ wheat} \quad\quad\quad\quad\quad = 500{,}000 \qquad\qquad\qquad\qquad 8{,}333\cdot3 \quad\quad 83\cdot3 \quad\quad 14$$
$$\text{Solonian} \quad \text{,,} \quad = \text{Χοῦνιξ of } gold \quad\quad\quad\; = 360{,}000 \qquad\qquad\qquad\qquad 6{,}000 \quad\quad\quad 60 \quad\quad 10$$

Χοῦνιξ or Kan of gold = Hand Cylinder = 360,000 grs. = 64 Hlf.-nail Cyl. or 64 Col. lbs. × $\dfrac{\pi o v s^3}{\text{foot}^3}$.

$$\left.\begin{array}{r}\text{Æg.}\\ \overline{\text{Eub.}}\\ \overline{\text{Sol.}}\end{array}\right\} = \left\{\begin{array}{l}120\\100\\72\end{array}\right. = \left\{\begin{array}{l}\dot{a}\mu\phi\text{ water}\\ \dot{a}\mu\phi\text{ wheat}\\ \text{Kan gold}\end{array}\right.$$

[a] This column is given for comparing the homologous Roman and Gothic measures, though the Greeks did not use this standard. See Table XI.

that quantity of alloy. Hence we may conclude that there was a Greek grain, and that it was the $\frac{1}{360,000}$ of the Solonian talent.

But that talent contained, as I have shown, 358,704 Troy grains. Hence the Greek grain was almost exactly equal to our modern grain. It is very slightly *smaller*. At first sight we should have expected to find it a little larger, about one-thirtieth.

But we must not forget that our modern Troy grain was arbitrarily enlarged by Henry VIII., and that it was previously $\frac{1}{17}$ smaller; and that before the introduction of the Colonia scale it was smaller still. We may safely say that the Greek grain was $\frac{358,704}{360,000}$ of the Troy grain, say ·9964. Thus the three talents contained (instead of the bewildering numbers in which I have hitherto expressed them in Troy grains), exactly

		Greek grs.
Æginetan talent		600,000,
Euboic	,,	500,000,
Solonian	,,	360,000.

Unless the Æginetan ὄβολος was divided into 16⅔ grains (an absurd supposition), we are compelled to suppose that the obol was itself the unit of weight—the lowest recognised sub-multiple of the talent—and that the grain was afterwards introduced by Solon, which is probable enough. After round numbers like 36,000 obols in a talent, 600 obols in a mina, 6 in a drachm, it is not likely or even credible that the Æginetan obol should itself be then divided into 16⅔ anythings. If it were divided into six or ten parts, each such part would be 2·76 or 1·6 Solonian grains, and Solon might possibly have altered this to suit the new scale; but it is more probable that Solon was the first to introduce the grain.

It will be convenient henceforth in speaking of the Greek weights to express them, not in Troy grains, but in Greek grains, merely bearing in mind that the latter is about ·9964 of the former, and correcting for the error when a very nice calculation is required. We now get the Greek measures in water, in wheat at Euboic ratio, and in gold, expressed in *Greek* grains, in the following table:—

CHAPTER IX

ROMAN MEASURES

SEEING that all the Roman bulk names are either the same as the Greek, or else translations of them (except the *modius*), we may well believe that at first they were the homologues of the Greek measures. Thus we have the *cochlear*, the *concha*, the *mystron* or *ligula*, the *cyathus*, the *acetabulum* (ὀξύβαφον), the *quartarius* (τέταρτον), the *cotyla*, the *sextarius*, the *congius* (χοῖνιξ), and the *amphora*. It is unnecessary to give an old Roman table of bulk measures and their water weights, because they are all the same as the Greek multiplied by $\frac{8}{9}$.

Ante-Sillian	Sextarii		*Post-Sillian*	Cubic inches
Ligula	$\frac{1}{32}$	$\frac{1}{36}$	Ligula	1
Cyathus	$\frac{1}{16}$	$\frac{1}{12}$	Cyathus	3
Acetabulum	$\frac{1}{8}$		,,	
Quartarius	$\frac{1}{4}$	$\frac{1}{4}$	Quartarius	9
Cotyla	$\frac{1}{2}$	$\frac{1}{2}$	Cotyla	18
Sextarius	1	1	Sextarius	36
Congius	2		,,	
,,	4	6	Congius	216
Semimodius	8	8	Semimodius	288
Modius	16	16	Modius	576
Urna	32		,,	
Amphora	64	48	Amphora	1,728
,,			,,	
,,			,,	
,,			,,	
Culeus	1,024	960	Culeus	34,560

SEXTARIUS

COTYLA

QUARTARIUS

LIGULA

CYATHUS

It would seem that at the time of the Sillian plebiscitum the pipe scale was already forgotten and lost sight of, though partly operative in practice, just as it now is in this country.

The principle on which the new *quadrantal* or cubic pes was subdivided seems to have been this. The cubical *amphora* was constructed. Then a cubical 6-inch *congius* was constructed and graduated in inches, like an ordinary glass or horn medicine-measure. When it was filled with water up to 1 inch, that measure was a *sextarius*, and this was taken as the *as*. It is a volume of water 1 inch high and 6 inches square; that is to say, it covers the floor of the *congius* to the depth of an inch. Here is a plan of the *sextarius*. Six of them one on top of the other make a *congius*, which is clearly the eighth of an *amphora*. Half a *sextarius* makes a *cotyla*, and a quarter is a *quartarius*, very naturally. But how are you going to halve the *quartarius*? You can take a third of it, as is seen by the plan: and this is just what the Romans did, and called it a *cyathus*. The *cyathus* then naturally splits up into 3 cubic inches or *ligulæ*. The whole mischief arose from dividing the foot into 12 *unciæ* instead of 16 *digiti*.

There is no doubt whatever that the Gothic standard wheat was of a specific gravity of ·8, that is to say $\frac{4}{5}$. And there seems to be little or no room for doubt that the Euboic ratio was $\frac{3}{4}$. To begin with, the Egyptian wheat of the Nile Valley was notoriously fine and heavy, and would weigh more than the wheat of the Northern countries. But quite apart from mere probabilities, the evidence is overwhelming that these were the Gothic and Euboic standards respectively.

I have no doubt that the continual necessity of comparing the weights of equal volumes of water and wheat was the cause of the introduction of the duodecimal scale.

We have now to ascertain which of these two standards was adopted by the Romans. I think we have only to look carefully at the *congius* to satisfy ourselves that the Gothic ratio prevailed at the date of the establishment of the *quadrantal*, and its eighth part the new *congius*. If the *congius* of water had been divided into 12 parts for the unit of weight, we should have known that the Euboic ratio was used. But ten *libræ* of water made a *congius*. Hence we may be certain that the *congius* held 8 *libræ* of standard wheat. And the ratio used was the Gothic.

What the *libra* was *after* the Sillian law we know exactly. It was the eightieth part of the weight of a cubic Roman foot of water. The water used was rain water, and the pressure and temperature were probably ordinary, and possibly neglected. This would make some uncertainty, but not very much. In Troy grains this is 4,989, with a possible error of 3 or 4 grains either way.

So far there is no difficulty. But what was the ante-Sillian *libra*? To begin with, we cannot be certain that the mint weight and market weight were the same. And even after the introduction of the cubical *amphora*, the coins hitherto examined do not bear out the supposition that the mint *libra* was 4,989. On the contrary, they point to a *libra* of about 5,050 Troy grains ; or somewhere between 5,040 (according to Hussey) and 5,054 (according to Böckh). There is good ground for believing that there were originally two weight units and weight systems in Rome, both of which were abolished or consolidated by the Sillian law : namely, the *pondus* and the *libra*. That the *pondus* was a specific weight is generally admitted, but it is commonly held to be merely another name for the *libra*. This is, I think, a mistake. We read also of the *dupondius*, or two-pound weight. Moreover, when used as the name of a specific weight and not a mere term for weight in the abstract, it seems to have belonged to the second declension. Instead of *pondere* we have *pondo*. Etymologically the word seems closely akin to our Saxon *pynd*, and there is good reason to suppose that it denoted the same weight—namely, the weight of a *sextarius* (or half-kan) of standard wheat. This would make it 6,614 Troy grains.

And what was the *libra*? It seems obvious that the *libra* with its twelfth part, or *uncia*, is the Etruscan λιτρα with its twelfth part, or ουγκια. But what was the λιτρα? According to the Tauromenian inscription, it was the $\frac{1}{120}$ of an Æginetan talent. But why should the $\frac{1}{120}$ part of a foreign measure succeed in driving out such a capital unit as the old *pondus*, and in establishing itself as the unit of weight for the whole Roman Empire? The answer is simple. We have seen how the Colonia pound or Tower pound, as it was called, superseded our good old Gothic pynd. And we have seen how Solon established a new talent which speedily eclipsed the two old talents, based as they were on wheat and on water. The answer now springs out without any dragging. The metal weight is bound to supersede the old measure weights. It is more handy, more durable, far more accurate, more portable—better in all respects. The λιτρα was a gold weight at first. It was the weight of a half-*cyathus* of gold, and therefore it weighed exactly 4,960 Troy grains. Now Solon's talent was a Greek kan of gold ($\chi o\iota\nu\iota\xi$) and his *mina* was a $\frac{1}{60}$ part thereof. But there is no reason to believe that the Oriental or sexagesimal scale was permanently established in Sicily. There is a sleepy Oriental want of precision about the *mina* scale which took no hold on the business-like energetic Western races. And the *mina* scale seems to have been rectified. We have seen what it was. 64 × 96 was, for convenience in counting by tale, smoothed down into 60 × 100 at a loss to the creditor of about sixpence in the £. The Italians, while adopting the Solonian unit, the kan-talent, divided it into 64 half-*cyathi* and took that weight of gold as their unit or λιτρα.

The λιτρα when it came in contact with the *pondus* would be found to be three-quarters of the latter to a grain. Hence they would work well together. It will be remembered that this is the precise relation subsisting between the Tower pound and our old Gothic pynd. History repeats itself. It will be seen that the λιτρα was not exactly a $\frac{1}{120}$ of an Æginetan talent—or, in other words, the $\frac{1}{100}$ of a Euboic talent—but very nearly, quite near enough to justify the statement as a guide to practice. Thus the λιτρα is 4,960 grains, and the $\frac{1}{100}$ of a Euboic talent 4,982 grains—a difference of only 22 grains, or about a pennyweight in a pound. When the *amphora quadrantal* took the place of the old *amphora*, the approximation became still closer: namely, 4,989 to 4,982. But this was a pure accident, and discloses no relationship whatever.

Concerning this *litra* there remains one more remark to make. Unlike Greece, Rome was a silver-using country. Greece used both gold and silver. And we have seen how Solon endeavoured successfully to couple the two together so as to make their exchange easy and convenient. It may be presumed that after the naturalisation of the λιτρα it would be represented chiefly, if not wholly, by its silver form.

What fineness of standard silver the Romans employed for this purpose we cannot say for certain : but we have two stubborn facts to guide us. Firstly, the Colonia standard silver was $\frac{3}{8}$ fine, and so remained for centuries. Secondly, the weight of pure silver in a *cyathus* of standard silver $\frac{3}{8}$ fine is 5,051 Troy grains, which is exactly the Roman money pound obtained from the actual weighing of Roman coins. Besides, $\frac{3}{8}$ is a natural Gothic ratio. It must not be supposed that when once it had been established for the purpose of the standard *weight*, there was any obligation or need to adhere to such standard fineness in the *coins* themselves. We know that this was not always done in England when the Tower pound was used.

The gold *libra*, though having all the appearance of a guess, is not in reality anything of the kind. We know that the Romans had the experience of Solon's legislation to guide them ; and what is more, we know that at a later date they actually did what amounts to the same thing a second time. They established in Colonia (Cologne) the famous half-*cyathus* gold pound—our Tower pound. Thus we are justified in tracing the history of the *libra* thus. It was *not* the *pondus*, which was probably the weight of a *sextarius* or pynd of standard wheat (Gothic ratio). It was three-quarters of a *pondus*. It was half a *cyathus* of pure gold = 4,960 Troy grains ; and it was exactly the weight of pure silver in a *cyathus* of standard silver $\frac{3}{8}$ fine, and was so established as the money pound. When the *amphora quadrantal* came in, the old *libra* was abolished for all purposes except the mint. The new *libra* or water pound was about 29 grains heavier than the old market *libra* of 4,960 Troy grains, which passed away into oblivion together with the *pondus*. And the explanation

GREEK AND OLD ROMAN PIPE SCALE

(The Roman weights are found by taking ⅔ths the homologous Greek weight)

Greek measures	Greek grains			Troy grains			Old Roman homologues
	Water weight	Wheat weight at ⅔	Gold weight	Water weight	Wheat weight		
Κοχλιαριον	36·6	30·5	703	36·5	30·4		Cochlear
Χμη	73·2	61	1,406·25	73	60·8		—
Κογχη	146·5	122	2,812·5	146	121·6		Concha
Μυστρον	293	244	5,625 g	292	243·2		Ligula
Κυαθος	586	488·25	11,250	584	486·5		Cyathus
'Οξυβαφον	1,172	976·5	22,500	1,168	973		Acetabulum
Τεταρτον	2,343·7	1,953	45,000	2,336	1,946		Quartarius
Κοτυλη	4,687·5	3,906·25	90,000	4,672	3,892		Cotyla
Ξεστης	9,375 a	7,812·5 c	180,000	9,344 a	7,784 c		Sextarius
Χουνξ	18,750	15,625	360,000 f	18,688	15,568		Congius
Χους	37,500	31,250	—	37,376	31,136		Chus
Ἡμιεκτον	75,000	62,500	—	74,752	62,272		Semimodius
'Εκτευς	150,000	125,000	—	149,504	124,544		Modius
	300,000	250,000	—	299,008	249,088		Urna
'Αμφορευς	600,000 b	500,000 d	—	598,016 b	498,176 d		Amphora

a Æginetan old mina.
b „ talent.
c Euboïc old mina.
d Euboïc talent.
f Solonian talent.
g λιτρα (Greek).

Æg. mina was afterwards 9,067 Troy grains.
Eub. mina was afterwards 6,303 Troy grains.
λιτρα (R. ligula gold) was 4,060 Troy grains.

of the accidental fact that the *libra* is $\frac{1}{100}$ of a Euboic talent, though a remarkable coincidence, is simply this: $\frac{1}{80} \times \frac{3}{4} \times \frac{8}{9} = \frac{1}{120}$. That is to say, $\frac{1}{80}$th of a *quadrantal* is $\frac{3}{4}$ of $\frac{1}{80}$ of an old *amphora* (which was 4 *modii* and not 3), and all Roman bulks being eight-ninths of the corresponding Greek bulks, we get the simple ratio, *libra* to Æginetan talent = 1 to 120. And seeing that the Euboic talent was five-sixths of the Æginetan, we get the still more remarkable ratio $\frac{libra}{\text{Euboic talent}} = \frac{1}{100}$, which may be arithmetically expressed thus: $\frac{1}{80} \times \frac{3}{4} \times \frac{8}{9} \times \frac{6}{5} = \frac{1}{100}$, a formula which explains the steps in the process. It was natural to suppose that the *libra* was derived direct from the ἀμφορεύς.

108   A SYSTEM OF MEASURES

CHAPTER X

FROM THE EARLIEST TIMES TO THE FRENCH REVOLUTION

WE have seen that the Gothic bulk-measures were all based on the yard. We have also seen that the Greek bulk-measures and those of the early Romans were precisely similar to our own, after correcting for the slight difference in the yard itself. The Greek name for the yard was δίπηχυς, and it was used by Herodotus and others; but in more recent writers it is supplanted by the πούς. The double yard was called an ὀργυια and the half-yard a πῆχυς. If we pass across to Asia, we find the Assyrians and Egyptians and Babylonians all making use of the some standard length measure, the cubit, and the same bulk measure, the bushel. We are justified in believing that the pipe scale, with its yard or half-yard for its unit of length, existed before the break-up of the Aryan family, and probably long before.

What was the yard? Or the half-yard, called the cubit? There is good reason to believe that it is of a far higher origin than is commonly supposed. Whether it has gained or lost in length since it left the hands of the great astronomer who constructed it is well worth inquiry.

We know that the Greek cubit (πῆχυς) was slightly longer than our own, and we know that the Roman cubit was slightly shorter. Between our own and the Greek length measures the ratio is 1 to 1·01125. Every hour the sun passes over fifteen degrees of the earth's surface. It passes over one-sixtieth of that distance per minute; and one-sixtieth of that again per second. What is the last-mentioned distance, expressed in cubits? Almost exactly 1,000 cubits. Allowing for the slight difference in the cubits we get:

Old cubits	per second	„	1,000
Greek cubits	„	„	1,003·01
English cubits	„	„	1,014·2

The Roman cubit has lost still more.

Actual measurements accord so well with tradition that it is difficult to dispute this origin of the sacred yard with any degree of assurance. I certainly do not propose to rake up the unprofitable controversy which has for years raged round the Great Pyramid. Much nonsense has been written on both sides. This much may be noted. The pyramid was admittedly built on mathematical and astronomical principles. It stands on a square base each side of which does as a fact measure

507·1 half-yards
= 501·5 πηχεες
= 500 old cubits.

If from these data we work out the circumference of the earth, we obtain a marvellously striking approximation—far nearer than that of Eratosthenes. It would therefore seem (in the absence of any other hypothesis) that in all human probability the original cubit was one thousandth part of the ground traversed by the sun in a second of time in the plane of the ecliptic, and that our sacred yard was originally the double of that.

The cubit (by whatever name known) was divided on the binary scale: the span was a half-cubit, the hand was a quarter-cubit, the nail was an eighth of a cubit, and so on. The double-cubit was a yard or διπηχυς; four cubits made a fathom or ὀργυια. Admitting the difficulty of preserving the cubit or any of its multiples and submultiples, especially in troublous times and among rude peoples, it would seem to be an even chance whether it would grow longer or shorter. As a fact, it appears to have grown shorter and shorter.

Even in Egypt there is reason to believe that it was at a very early period of its history slightly shorter than our own half-yard; but the evidence of this is somewhat shaky.

Of the Roman length measures before the adoption of the duodecimal scale we know little or nothing, except that the *pollex*, which was originally the sixteenth of a cubit, was probably degraded to the eighteenth in order to make it the *uncia*, or twelfth part of a foot.

Whether the *passus* was at one time 6 feet—that is to say, 4 cubits—we cannot say for certain, but it seems to have been regarded as the homologue of the Greek ὀργυια, which was four cubits. For land measures both peoples seem to have taken as a unit round numbers of the ὀργυια and *passus*, such as the σταδιον of 100 ὀργυιαι and the *millepassuum* of 1,000 *passus*. Whether our furlong was ever based on the yard or not I do not know. It seems more probable that it was independently chosen as the unit of land measure for sufficient reasons, and afterwards found to contain about 220 yards.

For bulk-measures, whether dry or liquid, there can be no reasonable doubt that cylinders were used *universally*, based on the cubit and its multiples and parts. We have seen what these are. The most popular, because the most commodious, of these was the bushel. The notion that the Greeks alone, while using the other measures of the pipe scale, adopted an anomalous measure of a bushel and a half for corn, and another of a bushel and an eighth for wine, must be rejected. The contention is *primâ facie* absurd, and I have shown how it arose. Allowing for slight differences in the length of the standard cubit, yard, and fathom, there can be no doubt that the *kikkar*, μεδιμνος, ἀμφορευς, old *amphora*, bushel, and firkin were all one and the same measure—namely, a cylinder 1 span in height and 1 span in radius.

It is equally certain that the standard weights were based on these cylinders of the pipe scale. The only question was, What is to be the medium? Agricultural people would naturally select grain of some kind. It is certain that they did choose wheat. Others, who had not regular access to grain, would as naturally select water as the medium. Hence we find all throughout the East, and also in Greece, that there were two scales of weight, one based on the weight of the water in the measures of the pipe scale and the other on the weight of the wheat. In Greece these two scales were known as the Æginetan and the Euboic. The latter does not appear to have any connection, as is usually supposed, with the island of Eubœa. The Euboic coins not only have an ox stamped on them, but one of these coins was actually called a βους. The term signified 'agricultural weight' as distinguished from 'water weight.'

For rough and rude transactions this grain weight was near enough and extremely convenient; but for nice operations it became necessary to establish a standard quality of wheat, and the evidence is overwhelming that the Greeks adopted the ratio of five-sixths, for obvious reasons, while it is equally certain that the Gothic peoples took four-fifths, for equally good reasons which I have explained. The Romans possibly used the Greek ratio in early times, but they clearly adopted the Gothic on the occasion of the great metrological revolution associated with the Sillian plebiscitum.

We have reason to believe that the weight of a bushel of water was the Æginetan ταλαντον, and that the weight of a bushel of wheat (Euboic ratio) was the Euboic ταλαντον. Then came Solon with his gold medium, but still using the pipe scale. And his talent is the weight of a χοινιξ of gold.

The Roman 'reform' consisted in substituting the cubic foot of water as the unit of bulk in place of the old *amphora*, and in substituting water for wheat as the medium. By so doing, the old pipe scale was utterly ruined and abolished. The adoption of the duodecimal scale clearly grew out of the Euboic ratio. People who were continually in the habit of comparing weights related to one another in the proportion of 5 to 6—that is, 10 to 12—are almost forced to adopt a duodecimal system. The individual who invented the *amphora quadrantal* must have been a man of no small genius. The difficulties he had to surmount were considerable. He could not touch the old popular measures, the *modius* or peck used by the peasants, and the *sextarius*, as it afterwards came to be called, the weight of which in wheat was the old *pondus*.

What he did was this: he divided the yard into 3 feet instead of 2 cubits. He then took a cubic foot instead of a cylinder, and he found that he could divide his cubic foot into 3 *modii* or pecks without perceptibly altering the *modius*.

Thus, $\frac{4\pi}{3}$ almost exactly equals $(\frac{1}{4})^3 \times \cdot 7854$.

By working upon this formula it will be seen that neither the *modius* nor the *sextarius* (the peck nor the pynd) were perceptibly affected. But the *congius* and several other measures less dear to the people were knocked out of recognition. We can only guess what the *congius* was by assuming that it was, as appears from the name, the homologue of the χοινιξ or kan.

This was the first and the last assault on the old pipe scale till we come to the *bouleversement* of 1789. But in the mean time, owing partly to 'the subtlety of merchants,' but more especially to the stupidity of legislators, the system itself had decayed and fallen into a condition of hopeless confusion. The very simple and beautiful principles on which it was based had been utterly forgotten and trampled upon. One eccentric Chancellor had gone so far as to propound a theory of his own concerning the true proportions of our liquid measures. What is almost more remarkable is that he succeeded in carrying his idiotic theory into practice, and to this day we are the victims of his whim.

Ignorant of the binary scale, he looked round for some reason why a pipe, a hogshead, a barrel, should contain so many gallons. Finding none, he hazarded a guess. No doubt, said he, they were intended to contain scores and units: thus a hogshead is three score and three, a tertian is four score and four, a butt is six score and six, and a tun is twelve score and twelve. This brilliant idea was embodied in a statute, and so farewell to the binary scale.

Similar misfortunes befell the system in all the other countries of Europe, and nothing was done till the great cataclysm of 1789, which gave us the metric system and the Code Napoléon.

PIPE SCALE
Elevation: full size

Kan. Hand Cylinder Congius. Χοινιξ

Gill. Nail Cylinder Quartarius. Τεταρτον

Half-nail Μυστρον

Qt.-nail Dram

8th Carat

Of this form were:—Pipe, culeus; boll; peck, modius, ἑκτευς; kan, congius, χοινιξ; gill, quartarius, τεταρτον; μυστρον; dram, κοχλιαριον; carat.

Pynd. Sextarius. Ξεστης

Acetabulum. 'Οξυβαφον

Skilling

Sceatta

Of this form were:—Quarter; bushel, amphora, ἀμφορευς; gallon, semimodius, ἡμιεκτον; pynd, sextarius, ξεστης; acetabulum, ὀξυβαφον; skilling; sceatta.

Gr.	
256	128th
	A 64th
128th	

A thirty-second

A sixteenth of a yard

Fl. drachm
Χημη
A 64th

Ounce
Cyathus
Κυαθος

A thirty-second

An eighth of a yard

(Kettle)
Mark Cotyla Κοτυλη

Sixteenth of a yard

A quarter of a yard

Stoup Χους

An eighth of a yard

Of this shape are:—tun; combe; tod, urna; stoup, χους; mark (kettle), cotyla, κοτυλη; ounce (ora), cyathus, κυαθος; drachm (fl.), χημη; double-carat; grain.

CHAPTER XI

HISTORY OF THE METRIC SYSTEM

AMONGST other rotten institutions which went into the melting pot in the French Revolution was the abominable system of weights and measures which, though nominally the same, varied in each province, and frequently also in many localities of the same province. There were more than fifty different livres; there were hundreds of different superficial measures of land, and a great variety of tuns for wine and other drinkables. In some places the length of the *aune* varied according to the textile fabric measured, and the weight of the pound according to the commodity sold. The attention of the Government, as well as of several *savants* in Paris, was drawn to the subject in consequence of numerous memorials presented by deputies to the States General convoked in 1789, and the National Assembly, in 1790, abolished all the feudal rights, including all those relating to the local regulations of weights and measures. Talleyrand was selected to prepare a report upon the subject. After referring to the advantages that would accrue to the land interest, to industry, and to the whole community from establishing uniformity in weights and measures, and after suggesting several systems based upon a natural unit as invariable as Nature herself, Talleyrand proposed to communicate with the British Parliament, with the view of obtaining their concurrence in the appointment of commissioners to be chosen in equal number from the Academy of Sciences in Paris and the Royal Society in London, to determine a natural unit of weights and measures. The first decided step taken in France for reforming the ancient system, and which eventually led to the establishment of the metric system, was consequent upon this report. This step was the issue of a decree of the National Assembly, dated May 8, 1790, and having for its object the determination of a natural standard unit for the basis of the new system. The decree refers only to a uniform international standard to be agreed upon by England and France, and to be derived from the pendulum. No mention is made of a standard of length to be derived from an arc of the meridian, or of a metric or decimal system. The English Government making no response to this appeal, the subject was referred to a commission of members of the French Academy, upon whose recommendation the ten-millionth part of a quadrant of the meridian was chosen by the decree of March 26, 1791, as the unit and base of the new system. The question was debated whether the reform of the weights and measures should be accompanied by a change in the system of arithmetical notation, by substituting a duodecimal for the decimal scale. The superior advantages of the duodecimal scale were stated and admitted, but they were alarmed at the practical difficulties of carrying out so great a change, which involved the addition of two ciphers to represent the numbers 10 or 11, as well as at the opposition it would be certain to meet with, not only in France, but in foreign countries; and they decided that it would be fatal to the adoption of their new system. On the other hand, they considered that nothing would so much facilitate its adoption as the use of the decimal scale, universally used in arithmetical notation, and they recommended it accordingly. A further decree of August 1, 1793, determined that the fundamental base of the new system should be called the metre, and that the metric system, with the decimal scale of division, should be adopted uniformly throughout the Republic.

The metric system was definitively established by the law of April 7, 1795. The construction or importation of any weights and measures of the old system was prohibited. Further laws and decrees followed for the more effectual carrying out of the new system, including one for the progressive substitution of the weights and measures of the new system in exchange for those of the old; the whole to be accomplished within two years, and at the expense of the State. All persons in business were to receive new metric weights and measures in exchange for their old ones. The date of this law was September 22, 1795. Then came the decree of November, 1898, for re-establishing offices of public weights and measures, in order to accelerate the introduction of the new system. The decree of April, 1799, relates to the verification and stamping of the new measures, and their exclusive use throughout France. About three months later a proclamation was issued regulating the standard metal of the new bulk measures. Meantime, the construction of the new standards, the platinum metre and kilogramme, was being completed, and they were deposited by the members of the Institute with the Legislative Body on June 22, 1799, when they were legally established as the primary standards of length and weight throughout France.

When it was determined, by the original Commission appointed by the French Academy, that the ten-millionth part of the distance from the Pole to the Equator should be the unit of length and be called a metre, they also decided upon the names to be given to its decimal parts and multiples. The tenth of a metre was called a *palme*, the hundredth was a *doigt*, the thousandth part was a *trait*. Ten metres made a *perche*, a hundred made a *stade*, a thousand a *mille*, ten thousand a *poste*, a hundred thousand a *degré*, and a million metres, or the tenth part of the quadrant of the meridian, was called a *décade*.

They also proposed to retain the old names of the weights; the tenth part of a gramme was called a *grain*, the gramme itself was called a *maille*, which is the name of an old French coin worth nearly a farthing; ten grammes made a *drame*, a hundred grammes made an *once*, a thousand grammes made a *livre*, ten thousand a *décal*, a hundred thousand a *quintal*, and a million a *millier*.

For the bulk measures, the old terms, *pinte*, *boisseau*, *setier*, and *tonneau* were retained. But, after the suppression of the French Academy in the revolutionary tumults of 1793, and the appointment of another scientific commission in 1795, the present names of the measures, in the decimal scale of which the parts are distinguished by Latin prefixes and the multiples by Greek prefixes, wer finally adopted by the National Convention, and they are annexed to the law of 1837, which is now in force. In 1812, however, Napoleon by decree permitted the use of the old weights and measures, provided they bore upon them their precise relation to the weights and measures of the metric system. This reactionary measure retarded the adoption of the new system in France, and from 1812 to 1837 the concurrent use of the two systems was sanctioned by law. In the latter year the Government of Louis Philippe passed the law of July 4, making the use of the metric system compulsory to the exclusion of all others throughout France from the first day of 1840.

The relation of the French monetary system of coinage to the weights of the metric system was considered essential to its completion.

The table of metric measures in Appendix I. is taken from the law of 1837, and is identical with that contained in the law of 1795. In conformity with the provisions of the latter law, each weight and measure of capacity in this scale is to have also its double and its half.

The States which have now completely adopted the metric system and made its use compulsory are France, Germany, Holland, Belgium, Italy, Spain, Portugal, Greece, Mexico, and the States of South America; it has also been rendered permissive in Great Britain and Ireland by the Act of 1864, and in the United States by the law of 1866. It has also been partly adopted in a damaged condition by Switzerland, Sweden, Denmark, and Austria.

As long ago as 1848 the Prussian Government entertained the project of submitting to the Parliament of the Germanic Confederation, assembled at Frankfort, a draft of law for the introduction of a common system of measures for the whole of Germany, and thus to supply a want which had been much felt since the establishment of the Zollverein in 1834. The Government appointed a special commis-

THE FRENCH MEASURES TABULATED ON THE BASIC SYSTEM

Length	Area	Bulk	Weight	Value[1]
Millimètre	...	...	Milligramme	
...	...	...	Centigramme	
...	...	...	Décigramme	
Centimètre	...	Millilitre	Gramme	
...	...	Centilitre	Dékagramme	
...	...	Décilitre	Hectogramme	
Décimètre	...	Litre	Kilogramme	
...	...	Décalitre	Myriagramme	
...	...	Hectolitre[2]	Quintal[3]	
Mètre	Centiare	Kilolitre[2]	Millier[3]	
...	...			
...	...			
Décamètre	Are			
...	...			
...	...			
Hectomètre	Hectare			
...	...			
...	...			
Kilomètre	Myriare			
...				
...				
Myriamètre				

[1] The French monetary unit, the franc, cannot be introduced into this table. It is 4½ grammes of silver, or rather 5 grammes of an alloy of silver and copper (9 parts silver to 1 part copper) called French standard silver.

[2] For measuring solids the *kilolitre* and *hectolitre* are called *stère* and *décistère*.

[3] The *quintal* and *millier* are a violation of the nomenclature rendered absolutely necessary by convenience. Most people call a *millier* a *tonne*.

sion, which came to the conclusion that a modified form of the metric system would be the best suited to the requirement of German trade, and they recommended the half-kilogramme as the unit of weight, and the Baden foot as the unit of length; but it was not till 1856 that the Prussian Government established the *Zoll-pfund*, equal to half a kilogramme, as the legal unit of weight throughout Prussia. Nearly all the Governments of the States composing the Zollverein followed this example, one by one, though not without making alterations in the sub-division of the *Zoll-pfund*. In June, 1860, the Diet convoked a commission of experts from the several States, who met at Frankfort in the following year, to discuss the question of introducing a uniform system throughout Germany. This commission, in its detailed report, recommended the French metre with its decimal sub-division, and it also proposed the adoption of the French nomenclature. The Berlin commission objected to any change in the standard measure of length, but admitted the importance of taking steps to get rid of the great diversity and confusion in the measures used throughout the different States of Germany, and of establishing a uniform standard. They further admitted that, having regard to the commercial dealings between Germany, Great Britain, and France, the choice lay only between the English and the French units. Both commissions agreed in peremptorily rejecting the English measures; firstly, because the English unit of length does not possess a simple relation to the unit of weight; secondly, because the English measures of length, area, and bulk are unco-ordinated; and thirdly, because the foreign trade of England with countries using the metric system is far more extensive than her trade with countries where the English system prevails. For these three reasons the Berlin commission agreed with the Frankfort commission that the metre and its decimal scale should be adopted in Germany.

Herr Brix, who drew up a report, strongly objected to the adoption of the French nomenclature, and pointed to the precedent of the Dutch Government, which legalised the metric system in Holland, with the decimal scale but without the French names, preferring the customary Dutch names for the new measures. The war with Denmark, in 1864, further postponed the settlement of the matter; but in 1865 Herr Brix, the Director of the Standards Commission at Berlin, furnished a report in favour of the metric system for Federal Germany. He still, however, maintained that the metre, though the best unit of length for international purposes and foreign trade, was too long for home use; and he also adhered to the German nomenclature. The war with Austria followed, and it was not till March, 1867, that the first decisive step was taken for introducing the metric system into the Confederation, and in June, 1868, a project of law was laid by the Prussian Government before the Federal Parliament.

The report of the committee of the Federal Parliament on this project of law is very interesting reading. Referring to the metre, it proceeds: 'As regards the French metric system, the mere fact of its original derivation from the length of the meridian does not constitute its chief value, for the accuracy of the measurements has not only been questioned but given up; it may, indeed, be fairly assumed that such a so-called natural standard will always be found defective, according as the progress of science enables us to measure and calculate more accurately from time to time. But, in fact, what from the first moment insured to the metric system, not only pre-eminence over all systems, but also its gradual adoption as a universal system, has been simply and solely the ingenious idea of assimilating its scale to that upon which our system of numeration is based. It is no conclusive objection that the number 10 can only once be divided by 2, without leaving fractions, and that when divided by 3 it gives no finite result. This defect, no doubt, is to be regretted, and the superiority of the number 12 may be readily acknowledged. So long, however, as the very important step of basing our whole system of numeration upon the number 12 remains unaccomplished—and probably it will never be so much as attempted—it is clearly right to base our entire system of measures, weights, and also coins on the number 10. However willingly we acknowledge the great step made by the present project of law, we cannot shut our eyes to the fact that the adoption of the *Zoll-pfund* as basis of the system of weights would seriously detract from the advantages expected from the new system. The consideration of this point may be left for discussion, while we confine ourselves at present to the expression of our convictions that the advantages of a strictly decimal system are so great as to render its adoption imperative upon us, even if no probability existed of other States following our example. But, as a matter of fact, the metric system is at this day in full working order in France, Holland, Belgium, Italy, Spain, Portugal, and Greece, while it is at least permissively introduced into England, Switzerland, and the United States, with every prospect of being eventually adopted in these countries. Thus the question can no longer be *whether*, but only *when*, the metric system shall be adopted by us. The reply is easy—immediately. Every day that we continue to use the old bad and diversified system is a loss to the national wealth. The new system once introduced, we should gain not only all its advantages in home trade, but also facilities that can hardly be overestimated in our foreign trade with metre-using countries. We have lost much by the delay which has taken place since 1861, but if the new system is to be adopted at all it must be adopted in its entirety. Half measures would necessarily lead to further changes and new annoyances to the public. The introduction of the *Zoll-pfund* may be held up as a warning. The great inconvenience and cost of its adoption will be within everyone's recollection. All was willingly borne because it was universally acknowledged to be a real step in advance, the new pound being exactly a half kilogramme. But if, instead of this half-step, the whole step had then been taken of adopting the kilogramme as our unit of weight, we should have been spared the present necessity of again making a change. To those who at the time advocated the adoption of the whole kilogramme it was pointed out that it differed too much from the unit of weight in common use, and that it was desirable to approximate the new pound as closely as possible to the old one. Now the truth lies just the other way. The transition from one weight to another is much easier when the new weight differs so much from the old as to preclude the possibility of confounding the one with the other. Thus the old pound of 28 loth has doubtless been often confounded with the new pound of 30 loth, but certainly never with the kilogramme of 60 loth. Nor will a new name be a drawback, but rather an advantage. For these reasons it is to be hoped that the opportunity once lost may now be retrieved. We cannot refrain from saying that it appears to us equally important, as forming the keystone of the whole edifice, to reconstruct the monetary system on the same principle as the system of weight and size measure.'

During the debate which followed objections were raised to the adoption of foreign names, difficult for the people to understand. Instead of introducing French names, it was proposed to use German. The answer was that the German people would learn to understand these names as easily as the French had done, for, after all, they were not French words, but were derived from the Greek and Latin. On the whole, it did not appear advisable to invent new names and to impose them by law. The Government project proposing the *Zollpfund* or half-kilo as the unit of weight was vigorously opposed by the committee, and rejected in the following amendment :—' The kilogramme is the unit of weight. The kilogramme is the weight of a litre of distilled water at its greatest density. The kilogramme is divided into a thousand grammes, with decimal sub-divisions. Ten grammes are called a decagramme, the tenth part of a gramme is called a decigramme; the hundredth part, a centigramme ; the thousandth part, a milligramme. The tun is a thousand kilogrammes.'

Another proposal of the committee was likewise adopted without discussion : 'To invite the President of the Confederation to submit as soon as possible to the Reichstag a new and strictly decimal monetary system, with a view to its offering the best possible guarantee of its extension to an international system.' All these recommendations were finally adopted and embodied in the law of June 13, 1868.

Less successful hitherto have been the efforts of the advocates of the metric system in the United States. Ever since the settlement of the constitution in 1789, in which it was declared that Congress has power to fix the standards of weights and measures, efforts have been made to establish a uniform decimal system. In 1790, in accordance with the recommendation of President Washington, Jefferson, then Secretary of State, reported elaborately on the subject, and recommended the seconds pendulum as the standard of measure. But in 1817 the matter was referred by the Senate to John Quincey Adams, who reported in 1821 advising a suspension of all innovation at home, until an international scheme could be adopted in America in conjunction with foreign nations; but no practical step for this object was taken till 1863, when the representatives from the United States, at the International Postal Congresses in Paris and Berlin, concurred in the resolutions passed at both meetings in favour of the general adoption of the metric system. Meantime, the National Academy of Sciences had been incorporated in 1863, and consisted of a body of not more than fifty scientific men, whose chief duty is, whenever called upon by any department of the Government, to investigate, examine, experiment, and report upon any subject of science or art. In the same year, the Academy appointed a committee on weights, measures, and coinage, including the most distinguished men of science in America. From 1863 to 1865 the civil war raged in America, but in 1866 the committee made their report, recommending that Congress should authorise and encourage the use of the metric system; and with this object made three practical suggestions: the immediate manufacture and distribution to the Custom Houses and States of metric standards; the introduction of the metric system into the Post Office, by fixing the maximum weight of a letter at 15 grammes instead of half an ounce, which is 14·17 grammes; to cause the new cent and two-cent piece to weigh 5 and 10 grammes respectively, their diameters to be made to bear a simple ratio to the metric unit of length. This report was referred to a Standing Committee of the House, with instructions to investigate the metric system and to frame a suitable Bill for its adoption by law.

Here are some extracts from the report of this committee : 'The metric system is already used in some arts and trades in this country, and it is specially adapted to the wants of others. It is, therefore, very important to legalise its use and give to the people the opportunity for its legal employment, while the knowledge of its characteristics will be thus diffused among men. Chambers of Commerce, Boards of Trade, manufacturing associations, and other voluntary societies and individuals will be induced to consider, and, in their discretion, to adopt its use. The interests of trade among a people so quick as ours to adopt a useful novelty will soon acquaint practical men with its convenience. When this is obtained—a period, it is hoped, not distant—a further Act of Congress can fix the date for its exclusive adoption as a legal system. At an earlier period it may safely be introduced into all public offices and for Government service. The nomenclature, simple as it is in theory and designed from its origin to be universal, can only become familiar by use. . . . After considering every argument for a change of nomenclature, your committee have come to the conclusion that any attempt to conform it to that in present use would lead to confusion, would violate the easily learnt order and simplicity of metric denomination, and would seriously interfere with that universality of system so essential to international and commercial convenience.'

These recommendations were embodied in the Act to authorise the use of the metric system of weights and measures passed by Congress on July 28, 1866. Congress also ordered the issue of a five-cent coin, weighing exactly 5 grammes, and measuring 2 centimetres in diameter. The following letter from the Hon. John Sherman, dated May, 1867, is worth preserving : 'I heartily sympathise with you and others in your efforts to secure the adoption of the metric system. The tendency of the age is to break down all needless restrictions upon social and commercial impulse. Prejudices disappear by contact. People of different nations learn to respect each other, as they find that their differences are the effect of social and local custom not founded upon good reasons. I trust that the Industrial Commission will enable the world to compute the value of all productions by the same standard, to measure by the same yard or metre, and weigh by the same scale. Such a result would be of greater value than the usual employments of diplomatists and statesmen.'

We have seen that when the metric system was first started in France, in 1789, the French Government tried to render the question an international one, and asked the co-operation of the British Legislature, but for some reason or other it met with no response in this country. Nine years later, after the completion of the triangulation and measurement of the bases for determining the new unit of measure, another attempt was made by the French to render the result of their labours useful for international purposes. Invitations were sent to the neutral and allied countries of Europe to send representatives to a Congress of scientific men to assist in the final settlement of a metrical system adapted to the convenience of all nations. This first International Congress on weights and measures met at Paris in 1798. It consisted of 26 members, of whom 14 were French and the remainder were Spanish, Italian, Danish, Swiss, and Dutch. England was not represented. The first important practical step towards the establishment of an international system was taken at the Great Exhibition in Paris in 1855, when an International Association was founded. Russia was the only great State not represented, owing to the fact that she was then at war with France and England, but she appointed a delegate to attend the fourth meeting of the Association in the year 1860 at Bradford, in Yorkshire. The Association formed a unanimous opinion that an international system must be based on the metric decimal system as established in France. The four fundamental principles agreed to by the committee in 1867 were :

1. That the decimal system, being in conformity with the system of numeration universally employed, is the most proper for expressing the multiples and parts of weights, measures, and coins.

2. That the metric system is perfectly fit to be universally adopted on account of the scientific principles upon which it is established, the homogeneity which exists in the relations of all its

parts, and the simplicity and facility of its application in the sciences and arts, in industry, and in commerce.

3. That the instruments of precision and methods employed for obtaining copies have attained such perfection that the accuracy of these copies meets every requirement of industry and commerce, and even the exigencies of modern science.

4. As every economy of labour, both material and intellectual, is equivalent to an actual increase of wealth, the adoption of the metric system, which may be ranked in the same order of ideas as tools and machines, railways, telegraphs, logarithms, &c., particularly commends itself from an economical point of view.

Great stress is laid by the International Association, and by all advocates of the metric system, on the saving of time in education which would result from its general adoption. The following is an extract from the memorial to the Chancellor of the Exchequer presented by the Association in 1859: 'Another most important consideration is the saving of time in education. Those who are employed in training boys for commercial pursuits bear unanimous testimony to the irksome labour of teaching the tables of weights and measures, and the system of arithmetic long established in this country, and they declare that the introduction of the metric system would be equivalent to the saving of at least a year to them and their pupils. This year is generally lost, because the weights and measures are for the most part forgotten as soon as learnt.' No doubt the advantages of the metric system are in great part due to its application of a uniform decimal scale, by means of which any amount can be expressed under a single denomination, thus simplifying computation and avoiding long and difficult reduction. The decimal system has been well described as a labour-saving machine.

The Weights and Measures Committee of 1862 was composed of members wholly favourable to the introduction of the metric system. They arrived at a unanimous conclusion that the best course is cautiously and steadily to introduce that system into this country, and they recommended that its use should be rendered legal. Unfortunately, they continued, 'No compulsory measures should be resorted to until they are sanctioned by the general conviction of the public.' They contended that the Government should sanction the use of the metric system together with our present one in the levying of the Customs duties; that the metric system should form one of the subjects in the Civil Service examinations; that the gramme should be used as a weight for foreign letters and books at the Post Office; that the Committee of Council on Education should require the metric system to be taught in all schools receiving grants of public money; that the public statistics of the country should be expressed in terms of the metric system in juxtaposition with those of our own; that in private bills before Parliament the use of the metric system should be allowed; and that the only weights and measures in use should be the metric and imperial, until the metric system has been generally adopted. As a consequence of this Report, a Bill to introduce the metric system was brought into Parliament in the following session, but it contained the provision that it should be made compulsory after the expiration of three years, and, although it was opposed by the Government, it passed the second reading by 110 to 75. The lateness of the session prevented the Bill from being proceeded with, but it was re-introduced in 1864 as a permissive Bill only, and finally passed as the Metric Act of 1864, since when it has remained, owing to the adoption of the permissive principle, just so much waste paper. Subsequent attempts have been made to make the system compulsory, but up to the present without success. The commissioners appointed to inquire into the whole question in 1868 observe that, considering the extent and importance of our commercial and scientific intercourse with so many nations who have adopted the metric system, and seeing that it has now become a widely extended international system, it appears that this country can no longer isolate herself from other countries and refuse to adopt that common method of computing the quantities of all merchandise and other articles passing between them which so many other nations have already accepted.

'With regard to the legal names to be given to metric weights and measures in this country,' writes Mr. Chisholm, 'the French nomenclature has already to a certain extent been adopted in the schedule to the Metric Act of 1864. Nor does there appear to be sufficient reason for making any alteration. These names are derived from Greek and Latin words, applicable to the decimal multiples and parts respectively. Although there are well-grounded objections to them on account of their inconvenient length and occasional similarity both to the eye and ear (for instance, decametre and decimetre), yet the general advantage of adopting in an international system not only uniform measures, but also uniform names, outweighs all such objections. The arguments in favour of one tell as much in favour of the other. If the metric system were generally accepted and used in this country, the names would probably soon be shortened. In France at the present time a kilogramme is generally called a kilo.' It is sufficient comment on this argument to observe that the very mistake contemplated by Mr. Chisholm as possible actually occurs twice in the Government Report itself. On page 138 *decigramme* occurs instead of *decagramme*, thus making the figures appear a hundred times smaller than they should; and on page 40 *decalitre* is printed instead of *decilitre*, this time making the quantity a hundred times larger than was intended. Mr. Chisholm proceeds to criticise the metric system in a somewhat hostile spirit as follows: 'The metric system has been described by its advocates as not only a scientific system, but as *the* scientific system of weights and measures. This is an assertion which will hardly bear the test of examination. A system of weights and measures to be used by the people must be a practical one. Judged by this rule, what is the metric system? First, as regards its units: the metre, the basis of the system, is declared to be a supremely scientific unit, based upon the length of the circumference of the earth itself, and ascertained by a scientific process to be the ten-millionth part of the meridian quadrant as determined by actual measurement. In the first place, no accurate measurement of the whole meridian quadrant is possible; secondly, it has been shown that the length of the quadrant of the meridian—that is to say, the distance of the equator from the pole—must vary in the different degrees of longitude. Thirdly, it does not appear that the slightest really practical advantage has been or could be obtained by establishing this relation between the metre and the meridian quadrant, even if it were actually and truly established, beyond the mere fact of determining a standard of length; and lastly, the metre is generally admitted not to be what it professes to be:—scientifically speaking, it is *not* the ten-millionth part of the meridian quadrant. In truth, the metre, as the unit or base of the metric system, has no other merit than that of being an established standard. It has no inherent scientific superiority over the yard, and for practical purposes it may be doubted whether its length is as convenient as that of the yard. The yard has been adopted in this country as the legal standard unit of length because it was practically found to be a convenient measure. It is the average length of the extremity of the outstretched arm from the mouth of a man. It is also a particularly convenient measure for women's use, as proved by the fact that it is the universal practice for milliners, dressmakers, and others to compute by it, and measure their materials for work by merely taking the length from the end of the second finger to the knuckle as an eighth of a yard, doubling it for a quarter, and so on. The yard has been in actual use in England as the measure of length from the time of the Saxon kings, and there has never been a question of its actual standard length. In the memory of man there has never been but one yard throughout England.

'Again, the metric system is described as a scientific system because of the simple and definite relation between the metre, its basis and unit of length, and the kilogramme and litre, the units of weight and capacity. But although this relation exists in theory, it is admitted that it has been found impossible practically to carry it out with scientific accuracy. The standard kilogramme is admitted not to be actually the weight of a cubic decimetre of pure water at the specified temperature, nor the litre a measure of capacity holding a cubic decimetre of pure water at its maximum density. Nor does much practical advantage seem to have been derived from the theoretical existence of this relation between the units of metric length, weight, and capacity. As regards the imperial system, there is no simple and definite relation, either theoretically or practically, between the units of length and weight, the yard and the pound; although the weight of a cubic inch of distilled water is specified in the Act which established the imperial system, and the relation between the cubic foot and the avoirdupois pound is also accurately defined. But the scientific relation between the unit of weight and of capacity—the pound and the gallon—is practically carried out, and constitutes the real test of the accuracy of imperial standard measures of capacity; whilst the cubical contents in inches of the gallon are defined by law.'

The reference to the milliners shows what straws the advocates of a system will cling to; for, as a fact, French milliners adopt precisely the same method for measuring their materials in terms of the metre; the length from the end of the second finger to the knuckle is just as often an eighth of a metre as it is an eighth of a yard; in both cases it is but a rough and ready method of obtaining an approximation. Again, it is quite as accurate to say that the kilogramme is actually the weight of a litre of water as it is to say that 10 pounds avoirdupois is the weight of a gallon of water—both statements are as nearly true as the precision of instrument-making admits of. To say that the yard has never varied in length from the time of the Saxon kings is a statement which requires a very powerful intellectual digestion. It has probably varied considerably more than the difference between the platinum metre at the Palais des Archives and the ten-millionth part of the meridian quadrant. But although Mr. Chisholm failed to appreciate the immense practical advantage of co-ordinating the several series of measures, he was obliged to admit the advantage which the metric system possesses over the imperial system in the simplicity of its decimal scale, as compared with the unscientific variety of scales in the imperial system; and yet he endeavours to whitewash even this crumbling mosaic by pointing out that it has been adopted because found practically adapted to the wants and habits of the English people. We have only to study the history and derivation of our measures, to discover that this statement is absolutely groundless. All other arguments failing, Mr. Chisholm falls back upon the difficulty of compelling the British public to substitute the new system for the old. 'You may, indeed, pass an Act of Parliament,' says he, ' to make the use of the metric system compulsory after a specified period, but does not the very term *compulsory* imply that the public will not willingly accept it ? Can you by any process of law practically enforce it against the wishes of a people so fixed in their habits and accustomed to self-government as the English people ? All history shows that in regard to weights and measures in this country the law has followed and confirmed the custom of the people, and that the habits of the people have not been dictated and enforced by the law.'

The successful introduction of the imperial gallon seems to have been overlooked. Again, it was the custom of the people to sell coal by measure until the law interfered and enforced its sale by weight. As for the difficulty of introducing a new system against the wishes of a public so hide-bound as the English, we have only to look across the seas to see how easily and rapidly the difficulty melts away in the face of a determined administration. There is no need whatever for any compulsion ; all that is necessary for the State to do is to decline to adjudicate contracts dealing with quantities unless expressed in terms of the system adopted. Outside the Courts there is no particular reason why people should not sell land by the stadium or verst, and wine by the hatful, if it so please them. It is only when disputes arise, and the litigants appear in Court to ask for State assistance, that the State very properly requires them to explain in intelligible language—that is to say, in terms of the legal system—precisely what quantities were intended by the terms made use of. Should the parties disagree as to the interpretation of these terms, the State is justified in declining to master any Volapük in which they may choose to express themselves.

CHAPTER XII

THE SYSTEM OF THE FUTURE

WHAT are the requirements of a really good system of measures? Firstly, it should be suitable for international use; secondly, the nomenclature should be clear and expressive; thirdly, the measures themselves should be exact; and fourthly, the system should be constructed with a view to economy.

Nomenclature.—Our nomenclature should be clear and expressive, and yet it should be simple. But it is possible to be too mathematically simple. Thus we might, if it pleased us, extend the simplicity of the metric system by calling the units of length, area, bulk, and weight respectively the mete, the metarea, the metebulk, and the metgrave. By so doing we should gain mathematical simplicity at the price of a good deal of clearness. The names of our units should be adapted to our varying needs. We want to measure the orbits of planets and the undulations of light, to weigh the sun and small gems; to survey a continent and a wire section. For the sake of economy our measure names should be short, and they should not be so similar as to occasion confusion. More important still, they should lend themselves easily to simple abbreviation.

From these three points of view it must be admitted the French nomenclature is not only bad, but the worst in existence. Words like kilometre and myriagramme are certainly not short, and words like decimetre and decametre are not only likely to lead to confusion, but actually do so. Nor is it easy to abbreviate names in which no fewer than eight begin with the letter 'm,' and several with 'milli' and 'deci' and 'kilo.' There is no more advantage in calling a thousand grammes a kilogramme, than there is in calling ten francs a decafranc. These simple relations are so quickly and so easily learnt that the cumbrous French nomenclature is an absurdly heavy price to pay for enabling classical scholars to perceive them at a glance. And what have the authorities to say on this question of nomenclature? In the first place, they are not agreed. Let us interrogate the Reports of our own and foreign commissions on weights and measures. Turning to the Report of 1869, we find the following: 'With regard to the legal names to be given to metric weights and measures in this country, the French nomenclature has already, to a certain extent, been adopted in the schedule to the Metric Act of 1864. Nor does there appear to be sufficient reason for making any alteration. These names are derived from Greek and Latin words applicable to the decimal multiples and parts respectively. Although there are well-grounded objections to them on account of their inconvenient length and occasional similarity, both to the eye and ear, yet the general advantage of adopting, in an international system, not only uniform weights and measures, but also uniform names, outweighs all such objections. The arguments in favour of one tell as much in favour of the other. If the metric system were generally accepted and used in this country, the names would probably soon be shortened.'

To pretend that the arguments in favour of a decimal system based on the metre tell as strongly in favour of the French nomenclature is the merest pretence. The arguments in favour of the latter are precisely those which might, with equal force, be adduced in favour of adopting the French language or Volapük. With equal force we might urge that, because we ought to endeavour to assimilate the monetary units of all civilised countries, therefore they should all of them be called by the same name—either we ought to call a sovereign a louis, or the French must learn to pronounce our word pound. But the whole contention is absurd on the face of it. The Commissioners seem to be of opinion that the main, if not the sole object of adopting the metric system is to secure international uniformity. If this were so, the adoption by other nations of our system and nomenclature would be equally efficacious. Those who take this view can have no conception of the immense economy, as a labour-saving machine, of a good system of measures, quite apart from its nomenclature. It must be added that throughout the whole of the ponderous Reports of our Royal Commissions, from 1819 to 1870, there is nowhere manifested any appreciation of these advantages, beyond a scant recognition of the gain of dispensing with compound arithmetic by arranging our measures on the decimal scale. As to a uniform international system, no one disputes its value for the purposes of commerce and scientific communication, quite apart from the local names given to the measures. It is difficult to overrate the benefits likely to accrue from such a world-wide scale. They would far outweigh even the advantage to be obtained by choosing the best unit, great as that advantage is. Even supposing the yard or the seconds-pendulum to be the best possible unit in theory, it would still be wiser to adopt the second best, or even the third best, everywhere, rather than to adopt one in England, another in France, and a third in America. Now the seconds-pendulum, though favoured by the first promoters of the decimal system in France, is nowhere in use, while the metre is used by more than half the continent of Europe, and the English yard over a still more extensive area. Hence, if it should come to choosing an international unit, the choice would probably fall upon one of these two. The question arises—which should be selected? It would assuredly narrow the area of change and upset to take a unit already widely used; and there would seem to be no reason (other than that of the dog in the manger) in compelling Englishmen to go through a period of temporary disorganisation for the mere gratification of French *amour propre*, nor, on the other hand, in putting France and the metre-using countries to the like trouble for the sake of British insular pride. We have, therefore, to choose between the yard and the metre on their intrinsic merits. We have seen what they are. The yard has nothing whatever to recommend it, while the metre is very simply related to the circumference of the earth, and is therefore very useful as an aid to the memory. Let us, therefore, choose the metre as our unit of length. It may be worth notice that the difference between the metre and the seconds-pendulum is less than a quarter of an English inch, the metre being about 39·37 inches and the pendulum about 39·14.

Having arrived at the conclusion that the requirements of a good system include precision, economy in time, in brain force, and in liability to error (which means double work), and fitness for universal adoption, we have now to inquire what principles are best calculated to attain these results. The metric system effects economy in two ways: firstly, by evading the necessity for what is called compound arithmetic; secondly, by minimising the difficulty of expressing one series of measures in terms of any other. By adopting decimal multiples and sub-multiples in all our tables of measures, we fit them into our current notation. Thus, if we have to add together several sums of money expressed in English pounds, shillings, and farthings, after adding up the farthings, we must divide by four and carry the quotient on to the pence column; then, after adding up the pence, we must divide by twelve and carry on the quotient to the shillings column; and, after totalling this, we must divide by twenty and carry the quotient on to the pounds column. Whereas, if we have to add together several sums of money expressed in French francs and centimes we do it by simple addition. Few would imagine what an immense saving of time and trouble this is when we come to deal with long and complicated accounts.

But even more important, from the point of view of economy, is the co-ordination of the several series. In our present English arrangement, the conversion of one series into another is a long, tedious, and often difficult process; in the new system it is done at sight. How many bushels go to a cubic yard? This should be a simple and easy question, and it is true that there are 'dodges' known to experts of obtaining a rough approximation; but to get an accurate result is lengthy and tedious. When a French child is asked, how many kilolitres go to a cubic metre, the answer is, a kilolitre is a cubic metre. To obtain an exact answer to the question, how many cubic yards are there in 743 bushels, we must proceed thus :—Turn the bushels into gallons, then multiply by $277\frac{1}{4}$ to turn it into cubic inches, and finally divide the whole by the cube of 36 to reduce it to cubic yards. Will the reader kindly work this out with his watch on the table, and note the time occupied? And yet this is a calculation which every farmer ought to make before building a barn, which should be big enough to hold the maximum yield of his land. Again, here is a square park of 100 acres; how long will it take to run round it at the rate of five yards per second? This is a complicated sum. Compare this: here is a square park of 100 hectares; how long will it take to run round it at the rate of five metres per second? Well, each side measures a kilometre, therefore the whole way round must be 4,000 metres; this gives us 800 seconds, or $13\frac{1}{3}$ minutes. But how many yards are there in one side of a square acre? Again, what space is required to stow away a ton of zinc, calling the specific gravity of zinc roughly seven? This is not so difficult as it would have been before the introduction of the imperial gallon, because that step co-ordinated our weight and volume series by making the gallon exactly 10 pounds of water. After learning that there are about $277\frac{1}{4}$ cubic inches in the imperial gallon—and, what is more, after learning to remember it—seeing that a ton contains 224 gallons of water, we multiply these two figures together, and so find the volume of a ton of water in cubic inches. After performing this multiplication, we divide by the cube of 36 in order to get it into yards, and then divide by 7 to allow for the specific gravity of the zinc; and there is the answer, somewhere about a fifth of a cubic yard. But if the question asked is, What is the volume of a millier of zinc? the answer is seen on the face of the question—one-seventh of a cubic metre.

In constructing a new system of measures for England, I propose to adopt the following half-dozen fundamentals: (1) the decimal scale; (2) the metre as the unit of length, (3) the co-ordination of the several series, (4) the basic system, (5) the abolition of units, (6) a system of abbreviation based on the chemical model.

1. *The Decimal System.*—There is no longer any hope of any other than the decimal notation being adopted for the ordinary purposes of numeration by the human race. That question was settled when our remote ancestors decided how many toes and fingers to wear. Seeing that this is so, no sane person sincerely pretends that any other scale of sub-division is better than the decimal. A decimal series of Troy ounces for bullion was legalised in 1853. And the Metric Weights and Measures Act was passed in 1864. In the words of the Report of the Committee of the Federal Parliament on a project of law for the regulation of weights and measures for the North German Confederation, dated 1868 : 'It is no conclusive objection that the number 10 can only once be divided by two without leaving fractions, and that when divided by three it gives no finite result. This defect, no doubt, is to be regretted, and the superiority in this respect of the number 12 may be readily acknowledged. So long, however, as the very important step of basing our whole system of numeration upon the number 12 remains unaccomplished—and probably it will never be so much as attempted—it is clearly right to base our entire system of measures, weights, and it may also be added moneys, on the number 10. The more the immense advantages are recognised resulting from the use of the decimal system in all the transactions of common life, as well as in the pursuits of science, the more important must it be deemed that these advantages should be secured to the fullest extent on the introduction of a new system of measures; and the more necessary it becomes that no more half steps should be taken.' An average French school-boy can, as a fact, do a given sum in French money in half the time required by the best mathematician in England to work it out in English money; the same is true of calculations dealing with weights and bulk measures. This question of a decimal scale stands apart by itself. For fifty years all our Commissions have pronounced in favour of the adoption of *some* decimal system. It is in no way affected by the disputes concerning the best unit or units, and concerning nomenclature.

2. *The Unit.*—One of the main objects in choosing this is that it

should be convenient for international purposes. Some have contended that it would be better, in face of international jealousies, to adopt a new unit. Others say, it would narrow the area of change and upset to take that which is already most widely used; and there is so much to be said for this view of the question that, even if the metre had no intrinsic merit, its adoption as the international unit of length would be reasonable.

3. *Co-ordination.*—There can be no doubt that from the point of view of economy in time and effort, it is an immense saving that the measures of surface, bulk, and weight should have some simple relation to the unit of length. Thus the hectare is a square area of 100 metres in length. But an acre is a square of something between 69 and 70 yards in length. Similarly, a litre is the cube of a decimetre, but a bushel is about 1·2837 cubic feet. If you want to know the dimensions of a reservoir to hold 64,000 kilolitres of water, you see at a glance that a reservoir measuring 40 metres each way would exactly hold it. But it takes a very complicated calculation to find the dimensions in yards of a reservoir capable of containing 64,000 tuns of water, whether expressed in gallons or in any other English measure of capacity. Even though this advantage were slight, which it is not, there is nothing to set against it. There is absolutely no advantage in having unco-ordinated measures. This principle, then, must be accepted.

4. *The Basic System.*—But if the object of co-ordinating the series is to furnish clear ideas of surface and bulk measures by at once calling up the length measure on which they are based, it is clear that surface measures and bulk measures should correspond with the length measures. Now the centiare does this; it is an area of one metre squared; and the are is an area of ten metres squared. But the decare does not fulfil this requirement; it is a square area of which the side or base is the square root of 1,000, a term which conveys no clear idea. Similarly, the deciare would be $\sqrt{10}$ metres, which conveys no clear idea; it is somewhere between 3 metres and $3\frac{1}{4}$ metres. Consequently, these measures have dropped out of the

French table. But when we come to the bulk series, we find equally unintelligible measures of capacity. The millilitre, the litre, and the kilolitre are cubic measures of which the bases are the centimetre, the decimetre, and the metre. But who can picture to himself a decalitre? It is a cube whose base is $\dfrac{\sqrt[3]{10}}{10}$, or 0·21544 metres. The like is true of a hectolitre, which is in common use as a wine measure, and it is true of all the rest of the series. They are all included in the French table, but have mostly dropped out of use in practice. It is clear they are not wanted, and might conveniently be abolished altogether. Anyhow, whether they are or are not useful as fractional measures in the ordinary transactions of trade, it is certain that they are not wanted as what may be called measures of account—that is to say, for purposes of calculation. The mere naming of measures renders them no more comprehensible than before, unless it enables the mind to grasp their precise meaning by furnishing at once their linear dimensions. The existence of such measures is a blemish on the French system, and is chiefly due to the unfortunate nomenclature introduced in the law of 1795, and finally annexed to the law of 1837. Let us revert to the rational method, and adopt the basic system, according to which the length measures rise by tens, the area measures by hundreds, and the bulk measures by thousands.

5. *Units.*—It will be seen that there is no longer any need for units of any kind except the standard unit of length, from which will be derived the other measures of length, and all measures of area, bulk, and weight; from which, again, will be derived all measures of value, pressure, force, power, temperature, heat, &c. Thus every measure in the tables may be regarded as a unit, or we may consider all units abolished, just as we please. It comes to the same thing. Of course the classical prefix nomenclature compelled the promoters of the metric system to fix upon a unit in each series. After some chopping and changing, they finally fixed upon the ten-metre squared as the unit of area measure, and they called it an are; for the unit of bulk, they chose the cubic decimetre, and called it a litre; and for

the unit of weight, they took the weight of a cubic centimetre of pure water, under certain conditions, and called it a gramme. The effects of this plan are deplorable, and we shall do well to take warning by the unfortunate French experience.

6. *Abbreviation.*—We have seen the difficulty, if not impossibility, of abbreviating the French measure names. Our own system, though rather better in this respect, is almost as bad as it can be. Hardly any of our measures can be symbolised by a single letter. Foot and furlong and fathom all begin with f, therefore foot is written ft. Between rod and rood it is difficult to distinguish. Neither gill nor gallon can appropriate the letter g; the like is true of pound and pennyweight; the latter is abbreviated dwt.—a mongrel combination of d for *denarius*, and wt., a contraction for weight. Pound is written lb., a contraction of the Latin *libra*. Hundredweight is written cwt.: c for *centum*, and wt. for weight. Pint and peck both begin with *p*. Ton and tun respectively denote a weight and a bulk-measure. And so forth. What we have to do in order to prepare for simple abbreviation is to give names to all the measures of length, area, bulk, and weight, such that no two names in the whole table begin with the same initial letter. For this purpose I have chosen the shortest available name, for the sake of economy, and in all cases monosyllabic words.

Upon these six fundamental principles I have constructed a system of measures suited to all purposes of trade and science, and in all respects fitted for universal use. I have also been guided by certain other considerations which may be regarded as less fundamental; for example, I have given names to the size measures which can easily be associated with familiar objects, and to the weight measures names which cannot be so associated. I have included in the system some measures considerably smaller than those in the French tables, and others considerably larger. I have avoided all names connoting number, such as quart, cent, mile, tierce, decade, and the like, as calculated to lead to false associations. Above all, I have abstained, after the unfortunate German experience, from using measure names which now stand, or have at one time stood, for approximately equal measures, thereby passing over such well-known terms as yard, mile, acre, pound, and gallon. Finally, I have kept quite distinct the coin names and their equivalent weight names, whereby some economic fallacies may be escaped. I have abbreviated some words and slightly altered the spelling of others in common use, but always with due regard to etymology. No word has been invented. And I have left room for supplementary fractional measures, if required, and suggested names for them—although there seems to have been little call for them in France, even for practical purposes. These fractional measures, as distinguished from measures of account, are not included in the table, and their abbreviations, if any, will consist of two letters.

The table on p. 141 will be found to conform strictly to the above six fundamental principles, and to the other conditions which I have imposed after a careful study of all past and foreign experiences.

To sum up, I have adopted the decimal scale for the sake of the immense economy of time and effort. I have adopted the metric standard because, although it is of small importance what unit we adopt, the selection of the metre is the most likely to secure uniformity throughout the world, and is in other respects the best in vogue. I have simply co-ordinated on the French model all the measures of size, weight, and value. I have introduced the basic system, as conducing to clearness of conception in the area and bulk series. I have abolished all the units in the derivative series, thus reverting to the rational mental process in which squares and cubes are realised and understood by reference to their bases. I have given names to the size measures which can easily be associated with familiar objects, and carefully abstained from doing so in the case of weight measures. I have given the 24 measures in the table names beginning with 24 different letters of the alphabet. Each name is monosyllabic, and, with the exception of hand and ton, they have never before been used for approximately equal measures. I have extended the bulk and weight series to far larger units than the French, and the area series to much smaller. And I have kept the coin names quite distinct from the weight names on which they are based. Each of the names in the table is represented by a single letter; and fractional measures, including coins of account, are represented by two letters, thus securing a nomenclature which is even more perfect than that of modern chemistry. Ask a modern chemist to revert to the ancient apothecary's nomenclature, and mark his reply. Language reacts upon thought, and a good nomenclature is to a science or an art what algebra is to quantitative reasoning. Its importance will be rendered more manifest when we come to the consideration of force measures, pressures, and measures of temperature.

CHAPTER XIII

THE BASIC SYSTEM

THE Basic System is simply the French enlarged and improved, and presented in an English garb.

I will now ask the reader to cast his eye over the names of the measures according to the proposed system. He can learn them off by heart in ten minutes; and if he will do so, he will be the better able to judge of their merits and defects after following the detailed explanation. At first sight, perhaps, they will appear somewhat strange and even grotesque, but a very short acquaintance will make them perfectly familiar.

Length	Area	Bulk	Weight	Value
Jot	En	Ove	Une	
Quil	Seal	Die	Gram	Cross
Hand	Foil	Litre	Yasp	
Mete	Nap	Vat	Ton	
Beam	Ar	Keep	Poid	
Course	Worth			
Reach	Ing			

In this table it must be borne in mind that the length measures rise by tens, the area measures by hundreds, and the bulk and weight measures by thousands.

From this it follows that the measures in the second or area column are the squares of the length measures on the same line, and the bulk measures in the third column are the cubes of the length measures on the same line, while the weight measures in the fourth column are the weights of pure water contained in the measures on the same line in the third column. The fifth column does not properly belong to the table, and it is inserted here merely as an adumbration showing the relation between the unit of value and the unit of weight, the former being one gram of pure gold; the whole subject of value and currency will be fully treated separately in the proper place.

Let us now explain these names one by one. The jot is the French millimetre. Ten jots make a quil—that is, a centimetre. Ten quils make a hand, which is a decimetre. Ten hands make a mete, which is the French metre. Ten metes make a beam, or decametre. Ten beams make a course, or hectometre; and ten courses make a reach, which is the French mile, or kilometre.

Coming to the next, or area column, the en is the square jot. A hundred ens make a seal, which is therefore a square quil. A hundred seals make a foil, which is therefore a square hand. For these three surface measures there are no French equivalents. A hundred foils make a nap, which is therefore a square mete. This is the French centiare. A hundred naps make an ar, which is therefore a square beam; it is the French are. A hundred ars make a worth, which is a square course, and is called in French a hectare. A hundred worths make an ing, which is therefore a square reach, and is called a myriare in French, though the term is not much used.

The third, or bulk column, begins with the ove, which is the cube of the jot. It is represented by no French equivalent. A thousand oves make a die, which is therefore a cubic quil, and is the French millilitre. A thousand dies make a litre, which is the French litre. It is therefore a cubic hand. A thousand litres make a vat, or cubic mete. This is the French kilolitre. A thousand vats make a keep, or cubic beam. For this there is no equivalent in the French system.

The fourth, or weight, column commences with the une; it is the weight of one ove of pure water at the temperature of melting ice in a vacuum. A thousand unes make a gram, which is therefore the weight of a die of water. It is the French gramme. A thousand grammes make a yasp, which is therefore the weight of a litre of water. It is the French kilogramme, and is popularly called a kilo. A thousand yasps make a ton, which is the weight of a vat of water. The French call it a millier, though the term is not uniform with the rest of the nomenclature of the metric system. A thousand tons make a poid, which is therefore the weight of a keep of water. It has no French equivalent.

This completes our list of co-ordinated measures. The whole table consists of but twenty-four words. Not two of these words begin with the same letter, and since there are only twenty-six letters in the English alphabet they are all used up but two. These two are x and z. This will be better seen at a glance by re-stating them in alphabetical order, thus :—

Ar	En	Ing	Mete	Quil	Une
Beam	Foil	Jot	Nap	Reach	Vat
Course	Gram	Keep	Ove	Seal	Worth
Die	Hand	Litre	Poid	Ton	Yasp

It will be observed that no figures are required in the table. The children will be taught, once for all, that the length measures rise by tens, the area measures by hundreds, and the bulk and weight measures by thousands. They will also be taught that each area measure is the square of the length measure on the same line, and that each bulk measure is its cube, and that each weight measure is the weight of the water contained in the bulk measure on the same line.

The unit of value is one gram of pure gold. It is merely a coin of account, and will not be actually coined; it is called a cross; it is abbreviated cr, and is also represented by some such symbol as the English £ or the American $. Ten grams of pure gold—that is to say, ten crosses—make a lion. This coin will contain exactly ten grams of pure gold and as much alloy as may be thought necessary for the sake of hardness—say one-tenth of its weight. Thus the actual coin will weigh 11 grams. It is worth 27s. 3½d. of our present money. The coins of account rise by tens. Ten doits make a groat. Ten groats make a cross. Ten crosses make a lion. More of this when we come to treat of the currency in detail.

To return to the bulk and weight series, it may be said that intermediate measures will be used in practice. If so, there is no objection to using them. It is a simple matter to give names to the tenths and hundredths of any one of the bulk or weight measures for the purposes of trade. Thus for medical purposes ten grams may be called a spoon, and ten spoons a gill. Then ten gills would be a yasp. Ten yasps may be called a lift, and ten lifts a sack. Thus we can retain our present sack as the weight of a tenth of a ton.

Small measures of bulk are not in request; but for larger fractional measures of capacity we may retain our present peck, with very slight alteration, as the measure of ten litres. Ten pecks may be called a cask, and then ten casks will make a vat. This gives us the following three little supplementary tables :—

Fractional Weights

Ten grams = One spoon,
Ten spoons = One gill,
Ten gills = One yasp,
Ten yasps = One lift,
Ten lifts = One sack,
Ten sacks = One ton.

Fractional Bulks

Ten litres = One peck,
Ten pecks = One cask,
Ten casks = One vat.

Coins

Ten doits = One groat,
Ten groats = One cross,
Ten crosses = One lion.

How long would it take an intelligent child of six to learn this table of co-ordinated measures and the three little supplementary tables of fractional measures and coins? And how long are the school children kept learning our present tables? It was said by one of our Royal Commissions that the adoption of the metric system would save a whole year in the education of our children. This seems to me an under-estimate. And when we have learnt our tables, how many of us contrive to remember them? Ask the first ten grown-up persons you meet in the street how many square yards there are in an acre, how many gallons there are in a hogshead, how many pounds there are in a ton, and even how many yards there are in a pole, and I shall be surprised if two out of the ten answer all four questions correctly and without hesitation.

But having once mastered the table of the basic system, the difficulty will be to forget it.

The method upon which I have chosen the particular names of the measures requires explanation. Almost any names would do, when once the mind has become accustomed to them. Thus some of our present names convey a meaning, while others convey none. A foot is the length of an ordinary man's foot; a hand is the width of his hand. But apart from association and experience, what idea does the word yard call up, or pound, or bushel? A furlong is the length of a furrow; but what is the length of the furrow? A mile is a thousand somethings; but what? Certainly not yards. Nevertheless, it seems better that each size name should either at once recall some familiar object of about that size, or should be easily capable of being associated with some such object. With the weight names, on the other hand, it is extremely undesirable that they should recall objects of known bulk, for thereby our ideas of specific gravity are blurred. We are apt to picture to ourselves a ton of gold as about the size of a ton of water; whereas it is about a twentieth of that size. Thoughtless persons are apt to think of a gallon as a weight of ten pounds, and to them a gallon of mercury would mean ten pounds of mercury, whereas such a measure would weigh over 135 pounds.

L

The next point is that names should never be invented and foisted upon the language; they should grow out of the language. Arbitrary sounds are detestable innovations, and will never take root. Imported foreign words should seldom be used so long as a good English word can be found.

With four exceptions I have tried to observe these rules. These four are mete, ar, litre, gram. For the sake of international convenience, I have accepted the French names for those measures which form the units of the metric system. I have slightly altered their spelling. All the rest are good English, and if they are not in all cases the very best that could be found, the difficulty must be borne in mind of beginning each name with a different letter. Take them one by one :—

Length Measures

Jot.—The smallest measure of length which it seems advisable to name is the French millimetre. It is about a twenty-fifth of our inch. If we look around for some common and familiar object of this length we have not far to go. We all read the papers, and we know what small print is. Many of us can tell at once whether the type is bourgeois or brevier; a smaller size still is called nonpareil. Now the letters a, c, e, m, &c. in nonpareil type each measure a millimetre in length. The letter which I have chosen to give its name to this measure of length is the Greek iota—an i without a dot—and the English for iota is, as everybody knows, jot. The word is already used both classically and colloquially to signify a very small measure. Thus we say, 'I don't care a jot.' 'Not one jot or tittle shall pass away.' I am satisfied that no name could be chosen to convey more definitely the idea of a millimetre. Its abbreviation is the single letter *j*.

Quil.—This is the name I have given to the centimetre, and it is perhaps the least satisfactory in the table. Abbreviation, *q*.

Hand.—This term is a familiar one to English ears. It was made a legal measure of four inches by Henry VIII. It is always used for measuring horses. It is within a sixteenth of an inch of our present hand, so that the substitution of the new hand for the old will not make the difference of an inch in the height of a big horse. And as it is not used as a measure in other branches of trade, there is no danger of its retention leading to confusion. Abbreviation, *h*.

Mete.—Having agreed to accept the names of the four French units, this term is already fixed for us. We have no occasion to go to the French or to the Greek for the word, as we have it already, in a slightly modified form, in our own language and with the same meaning. Mete-geard was a measure yard, or measuring staff, as long ago as Anglo-Saxon times. It is so used by Chaucer. Hence we are justified in dropping the letter *r* and reading mete. Abbreviation, *m*.

Beam.—For the purposes of association of ideas we do not require objects of a definite size. When houses are building we are accustomed to see beams of timber lying about, many of them the length of an average dining-room, say thirty-three feet. This is about the length we want a name for. Anyhow, it is quite as definite as our word pole, which might be a scaffold pole, a telegraph pole, a jumping pole, or a carriage pole. Abbreviation, *b*.

Course.—The Greek στάδιον was simply the length of the foot race course at the games. It was about two hundred yards. Every English school-boy knows pretty definitely what a hundred yards race is, quite as well as the Greek boys knew what a στάδιον was. And we cannot do better than follow the Greek example, and call our measure of about this length a course. Abbreviation, *c*.

Reach.—A reach is a long, straight piece of river or road; it may be roughly averaged at about a thousand metes. And in the absence of any more definite available object, it seems suitable enough to give a name to that measure. Abbreviation, *r*.

This completes the list of length measures. We see that a million jots make a reach. If anyone disputes it, let him get a million nonpareil iotas and put them end to end, and he will find that it is so.

Area Measures

The names of the area measures are the following :—

En.—A nonpareil en covers an area of about a square jot. Abbreviation, *e*. A hundred ens make a seal.

Seal.—If we look around for some familiar object measuring a square centimetre, we shall have a difficulty in finding one. But lawyers and persons who have occasion to transfer stock from time to time are familiar with those little bits of red paper which do duty nowadays for seals. You put your finger upon the little red wafer, and mutter 'dliver this smack indeed,' and virtue passes. Well, these little wafers are just about a square quil in area. And the term 'seal' seems well suited to denote this superficial measure. Abbreviation, *s*.

Foil.—If we take a hundred of these seals and paste them in ten rows of ten to the row, we shall find that they just about fill a page of a square duodecimo volume. In other words, the leaf of such a volume is just about four inches square. We cannot use the word leaf, because the letter *l* is required for litre, but we have the Norman English word foil, French *feuille*, which means the same thing. Abbreviation, *f*.

Nap.—A hundred foils arranged in ten rows of ten make a square mete, which I propose to call a nap. *Nappe* is French for a mat, and a small cloth, or nap, is still called a napkin. The term is easily associated with the size of an ordinary door-mat or the area of a square mete. Abbreviation, *n*.

Ar.—This term is fixed for us, for we have agreed to accept the names of the four French units. And the are is the French unit of surface measure. Abbreviation, *a*.

Worth.—Our old English word acre is the Roman *ager*, a field. Now a worth is plain English for a field, and we have but to accustom the mind to a square field one hundred metes long in order to have a perfectly clear conception of the area of the worth. Abbreviation, *w*.

FOIL, SEAL, AND EN, FULL-SIZE

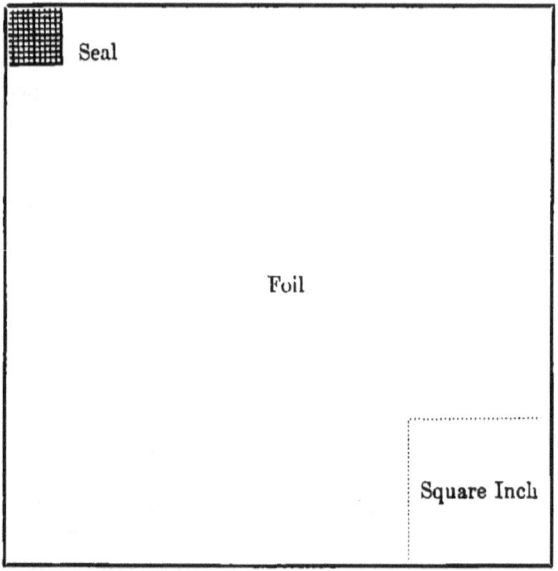

100 foils arranged in the form of a square (ten each way) make a nap, each side of which is a mete.

If a cube be placed on the seal as a base, we have the die; if a cube be placed on the foil as a base, we have the litre.

The large square is a foil, and its sides each measure one hand; the little black square is a seal, and its sides each measure a quil; the hundred very small squares are ens, and their sides are each one jot.

```
10 jots  = 1 quil     100 ens   = 1 seal
10 quils = 1 hand     100 seals = 1 foil
```

The dotted square in the right-hand bottom corner is our present square inch, which is put in merely for the purpose of comparison.

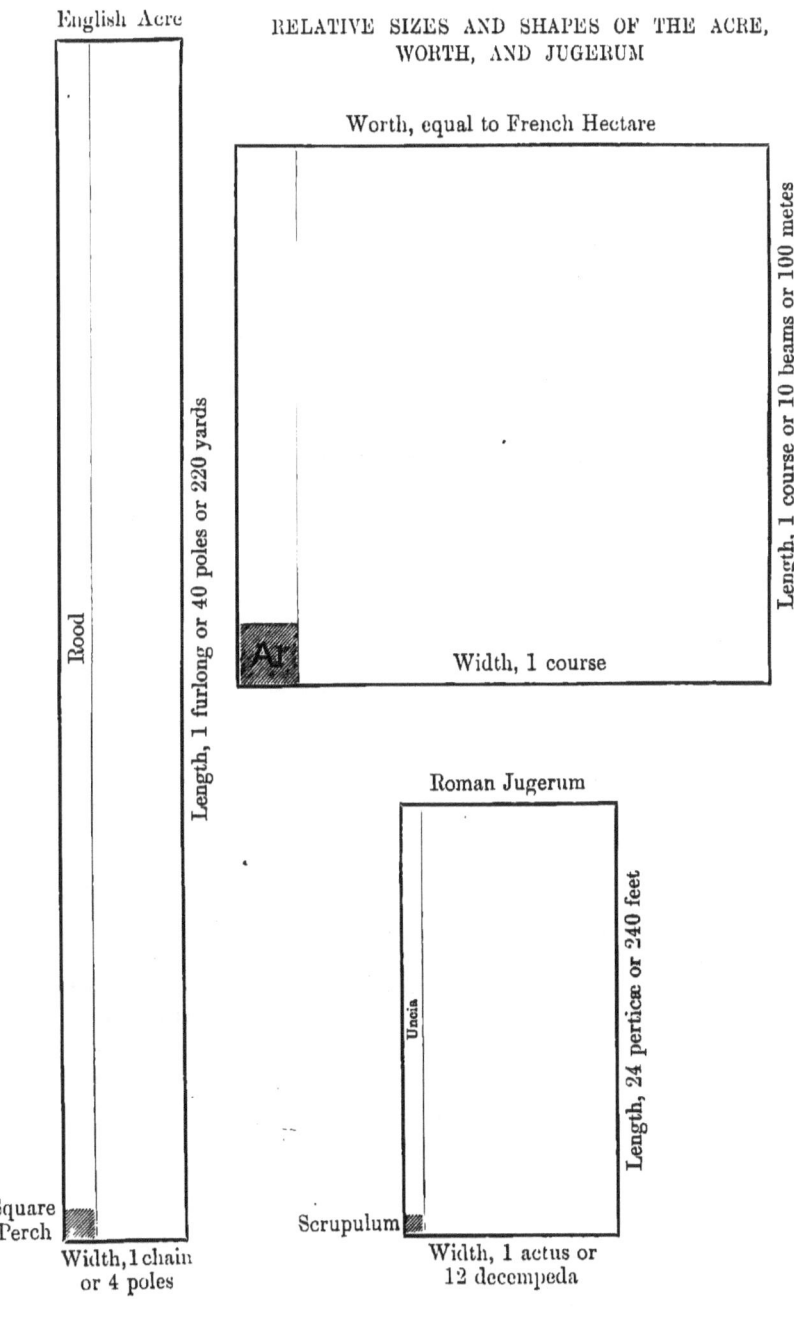

Ing.—It is difficult to find a suitable object in nature or art to give its name to so wide a surface measure as the square reach. Ley, would have suggested itself, but the letter *l* has been seized by the litre. The wide stretches of pasture near the banks of some rivers are known as ings, and for want of a more suitable term I have chosen ing to denote the area of a square reach. It is the French myriare. Abbreviation, *i*.

This completes the list of surface measures. It will be seen that if an ing were printed all over in nonpareil type without any spacing, there would be a billion letters in that ing.

Bulk Measures

This brings us to the names of the bulk measures, liquid or dry.

Ove.—An ove is a seed. The smallest seed of commerce is too large to give its name to this tiny measure of bulk. The measure we require is about a quarter of a spray millet seed, or a cubic jot. As a measure it is of little use, but it is well to name it for the sake of symmetry. Let us call it an ove. Abbreviation, *o*.

Die.—A thousand oves make a die. The French call it a millilitre. It is a cubic quil. If we look around for some familiar object which shall impress the size on the mind, we shall, after some search, hit upon the dice used in backgammon. French dice are a shade smaller than English, being usually just a cubic centimetre. This is precisely the bulk we have to name. Let it be called a die. Abbreviation, *d*.

Litre.—We are already pledged to the use of this term, for it is the French unit of bulk. It contains a thousand dies. Take ten dies and place them in a row, then take ten of such rows and place them in a layer, then lay ten of such layers one upon another, and you have a cube of one thousand dies. This is a litre, or cubic hand. It contains about a pint and three-quarters, and we think of it as a long drink. Abbreviation, *l*.

Vat.—A cubic tank one mete each way will contain exactly one thousand of such litres. It is the French kilolitre. Let us call it a vat. Abbreviation, *v*.

Keep.—If we take a thousand such vats, we shall, of course, find that they fill a cubic room or reservoir measuring ten metes each way. Certain reservoirs, kept at the sides of rivers for the preservation of fish, are called keeps; and I propose to confer the name on this the largest of our bulk measures. Abbreviation, *k*.

This completes the list of cubic or bulk measures, whether for liquid or for dry goods. It will be seen that there are a billion oves in a keep.

Weight Measures

We now come to the weight measures. As I have explained, it is not desirable to associate their names with any visible objects; and I have therefore chosen terms expressly designed to exclude the risk of any such association of ideas.

Une.—For practical purposes it is necessary to go much lower in the sub-division of weights than of bulk measures; and although the ove is likely to be of little use, the weight of an ove of water is required for scientific purposes, and is much used. The Board of Trade has thought it needful to procure a weight equal to the weight of half a cubic jot of water. The weight of a cubic jot, or ove, of water I propose to call a une. Abbreviation, *u*.

Gram.—Here we have no choice. We have agreed to accept the names of the four French units, and the name of their weight-unit is gramme, which has been shortened to gram. It is the weight of a die of water. It contains 1,000 unes. Abbreviation, *g*.

Yasp.—This is an old English word, signifying as much as one can pick up of grain, peas, or beans with the two hands put together in the form of a cup. Some old-fashioned grooms feed horses by the yasp instead of using the statutory corn measure. It may fairly represent a weight of about 2 English pounds, or 1,000 grams. It is the French kilogramme, or, as it is usually written, kilo. Abbreviation, *y*.

Ton.—A thousand yasps make a weight which differs from our present ton by less than a sixtieth part. We are, I think, justified in retaining the old name for this weight. It is one of the few popular weight measures in the country, and everybody has a pretty clear idea what is meant by it. There will be less difficulty in thus slightly altering the weight of the ton than there was in introducing the imperial gallon, and that step produced very little friction. Abbreviation, *t*.

Poid.—This term is merely a general name for weight. I propose to give it to the weight of 1,000 tons—that is to say, to the weight of a keep of water. It will be found to be a most useful measure. Abbreviation, *p*.

This completes the list of weight measures. It will be seen that there are a billion unes in a poid.

In comparing the new measures with the old ones, as an aid to the memory, we observe that the old and the new hand differ by less than a sixteenth of an inch. The mete is 1 yard 3 inches and a third, or thereabouts, and therefore its adoption would not much disturb our ingrained notion of the proper unit of length. In measuring a cricket pitch, three men out of four would be as likely to stride metes as yards. The beam is just about half our ordinary surveyor's chain; in other words, it differs from that measure by less than 2 inches and a third. The course is 109 yards and a third, and few could distinguish it by the eye from a measured hundred-yard foot-race course. The area measures are all of them unlike any of our present measures. The ing is just under two-fifths of a square mile. Coming to the bulk measures, the litre is about a pint and three-quarters; and 10 litres are slightly bigger than our ordinary peck. The vat contains 220 gallons, whereas our present ton weight of water measures 224 gallons. Looking at the weight series, the gram is a little more than half a dram. The yasp is rather better than 2 pounds. The ton is practically the old ton, or, quite accurately, it is 19 hundredweight, 76 pounds, and 10 ounces. For the precise relations between the old and new measures see the table on page 227..

One word in defence of the large units of bulk and weight which I have ventured to introduce, and which I think the promoters of the metric system would long ago have introduced had they not been cabined and bound by their unfortunate nomenclature. In dealing with road construction, canal building, water supply, and the like, our ton is far too small a unit. What we want to do is to train our minds to the clear conception of a cubic beam—that is to say, a room or reservoir ten metes high, ten long, and ten wide—and to give the conception a name. I have proposed the term 'keep.' Suppose we assume the quantity of water required to supply London at 75,000,000,000 gallons a year, is there any ordinary person, outside the ranks of the experts, who can form the vaguest conception of its bulk? Few persons have a very clear idea of a gallon, and when it comes to thousands of millions of gallons we might just as well tell them the number of raindrops or of teaspoonfuls. Express this same quantity of water in keeps, and we begin to get within the sphere of intelligible numbers. Three hundred thousand keeps is still a large figure. If we take the requisite supply per day at 250,000,000 gallons, we are still outside our reckoning, but when we call it 1,000 keeps per day we can form a perfectly clear and definite conception of its bulk. We picture to ourselves the ten-mete cube, or keep, and we construct in imagination a row or street of such keeps ten reaches long, and the thing is brought within our ken. Or, according to our bent, we picture to ourselves a cubic reservoir 100 metes deep, 100 long, and 100 wide, and again the conception is intelligible. I affirm that the keep is a useful unit of capacity for large quantities. By its aid we can compare masses of stone in a quarry, of earth in a tunnel, or of water in a canal without bewildering ourselves. For example, the quantity of water which passes daily over Teddington Weir averages 1,380,000,000 of gallons; this is, roughly, 5,400 keeps. The latter idea is intelligible; the former is not.

Moreover, we begin to understand and to compare these large quantities. We see that if the whole Thames flowed into the

THE BASIC SYSTEM

Companies' pipes it would fill them five times over. We see that water the depth of one beam over all the surface of the earth would just about equal the weight of the atmosphere. Again, if our unit of value is the gram of gold worth about 2s. 8¾d. of our present money, and ten grams make a lion, it is obvious that the pure gold in 100 lions will weigh exactly a yasp; and 100,000 lions will weigh a ton; and 100,000,000 lions will weigh exactly a poid. Our public debt in 1892, according to the *Times* of November 1 of that year, was about 900,000,000l., or, more exactly,

	£
National	689,944,026
Municipal	198,671,312
Total	888,615,338

Calling this 900,000,000l., we get, roughly, about 660,000,000 lions—that is, 6,600 tons, or 6·6 poids of pure gold. Taking the specific gravity of gold, roughly, at 20, we have 6·6 keeps divided by 20, or 330 vats—a most definite conception of bulk. Placed on a table one nap in area, it would stand 330 metes in height, which is 30 metes higher than the Eiffel Tower. The total quantity of gold used in the coinage of the world has been set at 950,000,000l., so that the payment of our public debt would use up nearly all the gold coins in the world. To this quantity of gold a like amount has been added as probably in existence for purposes of jewellery, gilding, and other ornamentation; and this gives us a total which would form a column on a nap base of about 725 metes high, or the height of a respectable mountain in the Lake District.

In 1890 our imports were 420,691,997l., which equals £308,200,000, or 3,082 tons of pure gold, which is rather more than three poids. Our exports were 328,252,118l., which equals £240,477,000, or 2,405 tons of pure gold. Difference, £67,723,000, which is 677 tons of gold. And the figures are brought easily within the compass of a ship's burden.

One ton of coal produces two-sevenths of a keep of gas, at a cost of just one lion; that is to say, 285·7 vats of gas. Coal makes 228 times its bulk of gas. One vat of coal weighs a ton and a quarter. If the gas companies of London consumed 2,000,000 vats of coal a year they ought to supply 456,000 keeps of gas. Perhaps an ordinary London house consumes about its own bulk of gas in a year.

Seven thousand million pounds passed the Clearing House in 1875. Expressed in lions as L5,134,000,000, we see at a glance that this sum in gold would weigh fifty-one and a third poids, and it would fill about two and a half keeps.

Let us now proceed to adapt the basic system of measures to the measurement of force, power or work, pressure, temperature, heat, specific gravity, and value; and we shall see how greatly simplified all such calculations are rendered by the process.

Unit of Force

The simplest method of measuring force is by weight lifted. And this involves two factors, the weight lifted and the height to which it is lifted. The English unit of force is the foot-pound. It signifies the quantity of force required to lift 1 pound a height of 1 foot. For the purposes of the basic system, we shall use the mete-yasp. This is, of course, the quantity of force required to lift a yasp the height of a mete. Since a mete is 3·2809 feet, and a yasp is 2·20462 pounds, we see that a mete-yasp equals 7·2331377 foot-pounds; or roughly, $7\frac{1}{4}$ foot-pounds.

The symbol for a mete-yasp is m–y; and if we have occasion to deal in large numbers of mete-yasps we may use the course-yasp, which is 100 mete-yasps, and the symbol of which is c–y.

Unit of Power or Work

We have seen that there are two factors in force. In power a third factor enters in—namely, time. The child which lifts a ton of bricks, one brick at a time, uses as much force as the elephant which lifts them all at once. But the power of the elephant is greater than that of the child. He does the work in far less time. In measuring *power*, therefore, we must compare the times in which the work is performed.

The unit of power used in this country is the horse-power, a technical term used to express the power required to lift a weight of 33,000 pounds a height of 1 foot in 1 minute This is 33,000 foot-pounds per minute. And this equals 4,424 m–y per min. But the figure 4,424 is an inconvenient and troublesome unit. Let us take 1,000 m–y per minute as our unit, and call it a power. Then 4·424 powers, or rather less than $4\frac{1}{2}$ powers, equals an English horse-power. We now understand that a 1 horse-power engine is capable of raising a weight of nearly $4\frac{1}{2}$ tons a height of 1 mete in 1 minute.

It is clear that we may use the symbol m–t per min. to express 1 power. Or if we please we may write 1,000 m–y, or 100 b–y, or 10 c–y, or simply 1 r–y per min. It would all depend on the figures we were dealing in. For it is clear that it would require the same power to lift a ton 1 mete as to lift a yasp 1 reach in equal times.

Unit of Pressure

Pressure means weight in proportion to surface. In England it is measured by the number of pounds to the square inch. It has been found by observation that the pressure of the air on the earth at the sea-level is on the average about 15 pounds to the square inch; but this is far from exact. In fact, the real mean pressure for the whole surface of the globe is probably a good deal nearer to $14\frac{1}{2}$ pounds than to 15. However, it is better to have a round figure as a unit of steam pressure than an intolerable fraction, and 15 pounds is the unit employed. Men of science are more particular, and for purposes of barometric pressure the round figure of 30 inches of mercury has been chosen to express the mean weight of the pressure of the atmosphere. Now this, though nearer to the truth than 15 pounds, is not accurate. The weight of 30 inches of mercury 1 square inch in section at 60° Fahr. is 14·67 pounds. The point on which I wish to insist is that if we are determined to take the weight of the atmosphere as our unit of pressure, we must be careful to choose some round figure which is an approximation to the truth, and this condition is fulfilled both by the practical engineer's 15 pounds and by the meteorologist's 30 inches of mercury. Anyone who wishes to be accurate can state what he holds to be the true weight of the atmosphere in terms of the conventional barometer. At the same time, it must be admitted that whereas the engineer's atmosphere gives us a very definite idea of the pressure of the steam, the barometer, on the other hand, gives us a very confused idea of the variations of barometric pressure. For instance, a rise of one-tenth of an inch means one-tenth of one thirtieth of 14·67 pounds to the square inch. It is easy to calculate the increase of pressure, but it is not clear on the face of it.

And now, what is to be the unit of pressure adopted in the basic scale? By all means let it be an atmosphere. And what is the mean weight of the air at sea level?

Let us assess the mean pressure at 14·22 pounds to the inch. This is as much below the truth as the engineer's atmosphere is above the truth. But it will suffice.

There are 2·20462 pounds in a yasp. This gives us a pressure of 6·45 yasps to the square inch.

And there are 6·45 seals to the square inch.

Hence we have the simple ratio, 1 yasp to 1 seal.

Surely this is a much more workable figure than 15 pounds to the square inch, and certainly far better than 14·67 pounds. It may be that the true mean weight of the atmosphere at sea-level will have to be expressed by 1·03, or by 1·029, according to the opinion of the meteorologist; but this signifies nothing.

We know that when the pressure of the atmosphere is at 1 yasp per seal, it will exactly balance a column of water 1 beam in height. And we see that when the column of water has risen to 1·001—that is, to a beam and a quil—the pressure of the air must be 1·001 yasps to the seal—that is, a yasp and a gram. In other words, for each extra gram of pressure the column will grow exactly 1 quil.

If, now, we find the water barometer too long for domestic purposes, we can substitute mercury on account of its great weight. We thus shorten our column 13·58 times. The precise length of the column is now of no consequence. As a matter of curiosity, it happens to be about 736 jots. It is no longer of any practical use to divide it into jots. What we want to know is the precise weight of the atmosphere above or below our unit of 1 yasp to the seal. We therefore divide our column into 100 or 200 parts or degrees, according to the description of barometer required, and each of these degrees will measure 10 grams of pressure.

If we adopt the form of barometer known as the Torricellian or J-tube, which is open at one end, and closed at the other, and of any uniform section (say one seal), then we must divide it into 200 parts. For we must not forget that for every half jot that the mercury rises in the closed tube it falls a half jot in the open end, so that the length of the column is increased by a whole jot. Hence an apparent rise of half a degree in the closed end indicates a total increase in the length of the column supported by the atmospheric pressure of 1 whole degree. And this increase is obviously 10 grams.

Each such degree, though this is immaterial, would measure 3·68 jots, and it can be sub-divided into halves or tenths, or into any number of parts suited to the purpose for which it is intended. And it would be as intelligible to the average reader as the ordinary pressure gauge.

To sum up, let us continue to adopt as our unit of pressure the weight of the atmosphere, or an approximation thereto. Let us call it an atmosphere. It is 1 yasp to the seal. And let us construct our barometers so that for every 10 grams of increased pressure the barometer rises 1 degree, instead of rising some fraction of an inch, which gives no quantitative idea of the increase without a complicated calculation.

Unit of Heat

The degrees marked on a thermometer tell us nothing beyond the fact that the temperature of a body is greater or less, and in what proportion. Thermometers tell us nothing whatever of the absolute quantity of heat in a body. The Centigrade takes the temperature of freezing water as zero, and the temperature of boiling water as 100°. But this division into 100 parts is arbitrary and means nothing. It merely tells us that 1° is the temperature of water the heat of which above melting ice is one hundredth of that which is necessary to raise it from freezing to boiling.

If the same quantity of heat be produced in an equal weight of another substance, say mercury, the temperature will be quite different. In the case of mercury, instead of 1° it will be 30°. Fahrenheit's thermometer is also graduated on a scale based on the melting- and boiling-points of water. But boiling-point is called 212°, and freezing-point 32°. Consequently, 9 Fahrenheit degrees go to 5 Centigrade.

Now, what we want is a thermometer which will tell us the force required to produce one degree of temperature. This has been called the mechanical equivalent of heat. In English figures the mechanical equivalent of the heat required to raise one pound of water 1° Fahr. is 772 foot-pounds. What we ought to do is to say that one degree of our thermometer shall express the temperature added to a pound of water by a mechanical force of 772 foot-pounds. We should then see the absurdity of using our present meaningless degrees. And we should get rid of the number 772, which is in no way interesting or convenient. We should take 100 or 1,000, or some other round figure, and base our degree on that. But we have nothing to do with feet or pounds. Let us agree that one degree of temperature according to the basic thermometer shall express the rise of temperature in a yasp of water produced by a mechanical force of 100 $m-y$. We find that it takes 235·298 $m-y$ to raise one yasp of water 1° Fahr. Hence 100 $m-y$ would raise it ·425° Fahr. This is rather less than half a degree Fahrenheit. It is ·236° Centigrade. Let it be our degree. And let us write it thus, 1ᶜ, with a small c instead of the usual little circle. This will avoid the necessity of explaining each time what thermometer we are using. Thus 1° Fahr. equals 2·35ᶜ, and 1° Centigrade equals 4·24ᶜ.

We see that water freezes at 0°, and boils at 424° at sea-level. But the force required to raise a given weight through a given height is precisely that force which is produced by the fall of such weight through the same space. If, therefore, 100 m–y will raise the temperature of one yasp of water 1°, the fall of one ton of water will raise the temperature of a ton of water 1°, and the fall of any volume of water whatever would suffice to raise such a volume of water 1°. Hence we may cancel the volume or the weight, and attend to the height of the fall alone. And we see that 1° is equivalent to 100 m, or shortly to c. Thus 27° means the temperature which would be produced by dropping any weight of water a height of 27 courses. But we must always remember that the medium in which our scale is expressed is water. Nobody will make the mistake of supposing that a vat always weighs a ton, simply because a vat of water weighs a ton; and similarly no one must make the mistake of supposing that if a yasp of lead fell a height of one course its temperature would be raised exactly 1°. As a matter of fact, the temperature of the lead would be raised 30°.

But since one ton, one yasp, or one gram would each be raised one basic degree by a fall of one course, it follows that though 1° is a sufficient symbol for a unit of temperature, it is altogether inadequate as a unit of heat. In order to get at this we must not omit to state the weight of the water. The heat generated by the fall of a ton of water is clearly 1,000 times that generated by the fall of a yasp of water, and this again is 1,000 times the heat generated by the fall of a gram of water.

For our unit of temperature we may use the symbol c, but for our unit of heat we must use the symbol c–y. And this is the quantity of heat generated by the fall of one yasp of water a height of one course.

Unit of Specific Gravity

We have agreed to use water as the measure of weight. We call any volume of water 1. Then whatever such a volume of any other substance weighs is its specific gravity. For solids and liquids we shall have no occasion to make any change in our tables of specific gravity. Mercury remains 13·58, gold is 19·2, &c. But when we come to the gases we find the present tables all topsy-turvy. Water is unthroned, and air is set up in its stead. And in some cases hydrogen is taken as the unit. But this puts us completely out of our reckonings. The reason of this is that long rows of decimals, beginning ·000, are unpleasing and confusing.

But in the basic system there is no such difficulty. We first fix upon our unit of volume. Let it be the vat. Then a vat of gold weighs 19·2 t, and 19·2 t is its specific gravity. And the weight of a vat of air is ·001208 of a ton. We have only to put a y instead of a t, and we at once avoid all these long rows of decimals. Instead of ·001208 t, we write 1·208 y, and the thing is done.

In our tables, the specific gravities of the gases will all be given in terms of water. Then, knowing the number of vats in a room—that is to say, its cubic measure—we shall be able at once to say what is the weight of the air in it, by multiplying 1·208 yasps by the number of vats. Thus, a keep of air when our barometer stands at the normal is 1,208 yasps.

For ordinary purposes it will be unnecessary to put either t or y after the number indicating the specific gravity of a substance, because it will be understood that when we are dealing with gases we mean yasps, and when we are dealing with solids or liquids we of course mean tons to the vat.

It is needless to say that the atomic weights of the elements and their compounds will in no way be affected by the introduction of the basic system of measures.

Value-measures

We have now to consider the best system of coins. The principle of nomenclature which I have adopted has been to connect them all with the weight of pure gold which they contain (if coined), or would contain in case they had to be coined, in gold. But I have in all cases carefully abstained from giving them the equivalent weight-

DESCRIPTION OF COINS

Double-lion	Double-cross	Double-groat	Double-doit
Lion	Cross	Groat	Doit
Half-lion	Half-cross	Half-groat	Half-doit

Double-lion: gold coin, weight, 22 grams; of which 20 grams pure gold, and 2 grams alloy; diameter, 32 jots; present value, 2*l*. 14*s*. 7*d*.; inscription, **2 lions, 20 grams pure gold, 1894.**

Lion: gold coin, weight, 11 grams; diameter, 28 jots; present value, 1*l*. 7*s*. 3½*d*.; inscription, **Lion, 10 grams pure gold, 1894.**

Half-lion: weight, 5·5 grams; diameter, 22 jots; value, 13*s*. 7¾*d*.; inscription, **Half-lion, 5 grams pure gold, 1894.**

Double-cross: gold coin, weight, 2·2 grams; diameter, 16 jots; value, 5*s*. 5½*d*.; inscription, **2 crosses, 2 grams pure gold, 1894.**

Cross: silver coin, weight, 27·5 grams; diameter, 38 jots; value, 2*s*. 8¾*d*.; inscription, **Cross, 25 grams pure silver, one-tenth of a lion, 1894.**

Half-cross: silver coin, weight, 13·75 grams; diameter, 34 jots; value, 1*s*. 4½*d*., nearly; inscription, **Half-cross, one-twentieth of a lion, 1894.**

Double-groat: silver coin, weight, 5·5 grams; diameter, 26 jots; value, 6·55*d*.; inscription, **Two groats, 5 grams pure silver, 1894.**

Groat: silver coin, weight, 2·75 grams; diameter, 20 jots; present value, 3·275*d*.; inscription, **Groat, one-tenth of a cross, 1894.**

Half-groat: silver coin, weight, 1·375 grams; diameter, 18 jots; value, 1·6375*d*.; inscription, **Half-groat, 1894.**

Half-groat: bronze coin, weight, 30 grams; diameter, 40 jots; value same as above; inscription, **Half-groat, 5 doits, 1894.**

Double-doit: bronze coin, weight, 12 grams; diameter, 36 jots; value, ·655*d*.; inscription, **2 doits, 1894.**

Doit: bronze coin, weight, 6 grams; diameter, 30 jots; value, ·3275*d*.; inscription, **Doit, 1894.**

Half-doit: bronze coin, weight, 3 grams; diameter, 24 jots; value, ·16375*d*.: inscription, **Half-doit, 1894.**

Counting the two half-groats, this gives us thirteen coins of exchange, the diameters of which progress by 2 jots at a time from 16 jots for the gold double-cross to 40 jots for the bronze half-groat.

COINS IN ORDER OF DIAMETER

Name	Diameter in jots	Weight in grains	Present value		
Double-cross	16 gold	2·2		5s.	3¼d.
Half-groat	18 silver	1·375			1·6375d.
Groat	20 silver	2·75			3·275d.
Half-lion	22 gold	5·5		13s.	7·75d.
Half-doit	24 bronze	3			·16375d.
Double-groat	26 silver	5·5			6·55d.
Lion	28 gold	11	1l.	7s.	3·5d.
Doit	30 bronze	6			·3275d.
Double-lion	32 gold	22	2l.	14s.	7d.
Half-cross	34 silver	13·75		1s.	4·375d.
Double-doit	36 bronze	12			·655d.
Cross	38 silver	27·5		2s.	8·75d.
Half-groat br.	40 bronze	30			1·6375d.

Of these coins, three will be very useful as handy measures, the bronze half-groat, which is exactly 4 quils; the doit, also bronze, which is exactly 3 quils; the groat, silver, plentiful, exactly 2 quils. The rest would as practical measures be of little or no use. But it is of the greatest importance that we should be able to distinguish the several coins in the dark by feeling their size. A new farthing may easily be mistaken for a half-sovereign in a bad light; and when it is quite dark, half-sovereigns are frequently given to cabmen by mistake for sixpences. In the system here proposed, there are no two coins which do not differ by at least 2 jots in diameter, whereas our present florin, penny, and half-crown are about 30, 31, and 32 jots respectively; and the half-sovereign, the sixpence, and the farthing are all within 1 jot of each other.

The thickness of the coins will also vary somewhat more than they do at present. And of course there is no reason against milling the gold and silver coins which does not avail now.

The area of a coin is limited. Whatever it is right that the public should know should have the first claim to space, after which whatever room is left may fairly be filled in with the devices of art and tradition.

The weight of pure gold in the gold coins should be stated thereon, and the weight of pure silver in the silver coins should also be stated on the face of the coins. And if the ratio of gold to silver coins is guaranteed, then the guaranteed ratio should appear on the face of the coins. If all this takes up room, we might dispense with Fid. Def. and D.G., and some of the ancient titles of the reigning monarch.

THE BASIC SYSTEM

name. Coins may be divided into coins of account and coins of exchange. We make our accounts in terms of £ s. d., and not in terms of half-crowns, threepenny-bits, and halfpence. Yet it is convenient to have fractional coins for the purpose of easy exchange.

Let 1 gram of pure gold be called a cross. It is merely a coin of account, and will not be actually coined in gold. Let 10 grams of pure gold be called a lion. Lion is as good a name as eagle any day, and certainly much better than sovereign or pound. To these 10 grams of gold 1 gram of alloy is added, so that the actual lion-piece itself weighs 11 grams. The value of the new cross is 2s. 8¾d., or rather more than a modern half-crown. The tenth of this is called a groat, and is worth just over threepence farthing, or 3·275d. The tenth of this, again, is called a doit. It is worth about a third of a penny.

These four—the doit, groat, cross, and lion—are all that are necessary for purposes of account. For fractional exchanges their number is increased to thirteen, viz.:—

1. Double-lion 4. Double-cross 7. Double-groat 11. Double doit
2. Lion   5. Cross   8. Groat   12. Doit
3. Half-lion   6. Half-cross 9 & 10. Half-groat[1] 13. Half-doit

Of these, the first four are of gold, the next four of silver, and the last four of bronze. The cross actually used will be a silver coin containing 25 grams of pure silver and 2½ grams of alloy. It will bear on its face the number of grams of silver which it contains; it will be a silver coin, and not a token coin, like our present crown, which it will resemble in size and weight. The State will, through banks of repute and stability, guarantee the holders of these coins for ten years at the rate of ten to the lion. And, in a similar way, it will guarantee the holders of the gold coins at the same ratio in silver.

[1] The half-groat will be minted both in silver and in bronze, and although it is probable that one of these coins will drive the other out of circulation, it is well to give the public the option. I myself incline to believe that the large bronze coin would be very popular. Most people think otherwise.

CHAPTER XIV

MONEY

WHEN the immense importance of co-ordinating the measures of length, area, bulk, and weight has been recognised by the public, they will not long remain blind to the necessity for similarly co-ordinating the measures of value with those of weight; and when this time arrives the absurd reverence for our English pound sterling—'the fixed star of our monetary system'—will also pass away; and we shall recognise the folly and provincialism of asking other nations to join us in establishing an international currency based on a unit so arbitrary, meaningless, and inconvenient. Can any human being explain the mystic virtue of our British pound sterling, of 123·27448 grains of standard gold, $1\frac{1}{2}$ fine? The lull in the agitation on behalf of the metric system in this country is probably for the most part due to the vulgar fear of disturbing our existing coinage system, if indeed the word 'system' is applicable in this connection. The decimalisation of our coinage is doubtless much to be desired, but it is of at least equal importance that we should adopt as our unit of value some definite weight of pure gold, likely to approve itself to other nations. If we adopt the metre as our unit of length, we must adopt the gramme of gold as our unit of value, so long as we adhere to the single gold standard. No step could be more mischievous at the present time than to introduce a decimal system of coinage on the pound-and-mil system. Other nations would never accept it, and sooner or later, probably later, we should find ourselves in the unenviable predicament of the German Government after their unfortunate adoption of the *Zollpfund*; in other words, we should be compelled to make a clean sweep of our new coinage and start afresh on a rational basis. It is not likely that the importance of this consideration will be made manifest to the English public until they have accustomed themselves to the use of the gramme and its multiples, and it is for this reason that the decimalisation of the coinage might wisely be postponed until after the adoption of an exclusive metric system of measures of size and weight. This is not the view taken by the International Coinage Commission of 1868, appointed in consequence of the proceedings at the International Monetary Conference of Paris in 1867. They stated their opinion that while some advantage might be derived from a system of international coins, the full advantage anticipated could only be attained by a general system of international currency, and they entertained no doubt that these advantages would be still further increased by an assimilation of weights and measures; but they added that they did not consider it necessary that any measure for the assimilation of the currencies of the principal nations of the world should be postponed until steps are also taken for the assimilation of weights and measures. At the same time, it is difficult to see how a rational standard of value could be acceptable to the people of this country before they have made themselves acquainted with a rational system of measures on which to base it. It is for this reason that I cannot for one moment concur with the Commissioners of 1868 when they say, 'We are of opinion that even if the difficulties of establishing an international unit of coinage cannot be at present overcome, yet the decimalisation of *our* system of coinage, which is in the power of the Government, would be very useful to the public;' and I can concur with them heartily in the opinion they express in what seems very like a flat contradiction of the preceding statement: 'We are of opinion that it is expedient that no legislation should take place with respect to the metric system until the whole subject of the weights and measures of this kingdom be brought before Parliament in one Bill.'

Money is a standing testimony to man's dishonesty. Exchanges can be conducted on two systems—immediate barter and credit. Money is a form of immediate barter. If all men were always honest, money would not be required. A. is a hatter, B. is a dairyman. B. wants a hat worth 50 quarts of milk. A. agrees to the exchange. Now, if B. would supply him with 5 quarts a day for 10 days all would be well. B. promises to do so, but A. does not trust him. Hence, instead of milk, B. must hand over at once, on receipt of the hat, something of equal value, which can be returned by him in instalments in exchange for milk. The most convenient, handiest, most durable, divisible, and assayable thing to be found is some precious metal. Unlike precious stones, it can be indefinitely subdivided without losing proportionate value. Cut a ruby in two, and the two halves are worth much less than the whole. Cut a bar of gold in two, and the two halves are exactly worth the whole. Gold, though not of everyday use for ordinary purposes, like lead or iron, has these advantages: it is far more compact in proportion to its value and can be carried about more conveniently; it does not oxidise or wear away as fast as lead; it is cleaner to handle; it is of a distinctive colour, and it is of a distinctive weight; and above all, owing to the fairly constant demand for it for particular uses, its value is fairly constant. It is probably impossible to find any commodity whose value varies within narrower limits over a considerable period than gold. Anyhow, gold has now become in most civilised countries the recognised medium of immediate barter. Silver, copper, nickel, and some alloys are also used for smaller exchanges; but where the values to be exchanged are large enough, gold is, as a rule, preferred as the medium. This is a fact, quite apart from the currency twaddle. What is to be borne in mind is that money is a barter medium, and has no connection whatever with credit. The better the credit of the citizens of a country among themselves, the less the amount of the barter medium required for carrying out exchanges. If the dairyman's promise, either verbal or in milk notes, is as good as his gold, it is clear that the carrying, storing, and exchanging of the heavy yellow metal is so much waste labour.

England uses less barter metals than any other country in proportion to the amount of the exchanges effected. There is nothing magical in gold. Legal tender is only legal fraud, or else legal foolery. If a man contracts with another to pay a ton of pig-iron, it is unjust to let him pay an ounce of gold. If he contracts to pay an ounce of gold, he is morally bound to pay it, whether it is legal tender or not. Surely men are not such infants as to forget to state what they are agreeing to exchange. 'Here are ten quarters of wheat for sixteen pounds,' says A. B. accepts the wheat, and straightway the State steps in and says, 'Mind, now, by sixteen pounds you meant, or must be taken to have meant, so much gold with our stamp upon it, and containing so much rubbish.' If the State would only get out of the way and mind its own business, A. would ask B. what he meant by sixteen pounds, and he would receive a plain answer.

We hear a good deal about standards and proper standards of value, but is there any necessity for the State to trouble itself about standards at all? I confess I cannot see any. It may be an interesting problem for plutologists to discover a fairly uniform measure of value, by means of which prices at different dates may be quantitatively compared. For it is of no use to tell us that wheat was fifty shillings one year and thirty-five another, unless we are sure that the value of the shilling or unit is the same in both cases. That is perfectly true, but it by no means follows that the State is competent to discover this standard. Nor does it follow that it would be of any practical service to trade if discovered.

Instead of two standards, or a double standard, or five standards, the proper course for the State is to abolish standards altogether. If A. delivers goods to B. on the promise of B. to pay an agreed weight of gold at some future date, it is A.'s and B.'s look-out only. If the value of the given weight of gold varies between the dates, so much the better for one and so much the worse for the other. It is no affair of the State's. It may be said that A. and B. are thus compelled to speculate in a metal in which they have no wish to deal. This is so. It must be so in the nature of things. But gold is about the least speculative commodity to deal in ; the gains and losses are very slight, and, furthermore, they tend to cancel one another on a number of transactions.

There has been so much nonsense spoken and written about money that it would be a good preliminary step to abolish it altogether—in thought, if not in practice. A pound or sovereign means, or should mean, such a weight of gold. Whether it is bought and sold in the form of small medals or cakes, stamped or unstamped, is of no consequence. Then why these little medals? Clearly, if gold were not made into coins it would be necessary to weigh and test the gold, just as it is now necessary to weigh and test iron or cotton before dealing. This is a tedious process, and doubtless the private individual who hit on the happy expedient of running the metal into little cakes or bars of equal size, weight, and purity, and marked with the same, would be hailed as a public benefactor. As a guarantee of genuineness he would stamp his own name on the coins. A firm of great wealth and repute would soon obtain a good hold on the market.

Coins are still extant in this country bearing the names and crests of private individuals and not guaranteed by the State. At the beginning of the seventeenth century the practice of private minting was extensively carried on. It is true that the monopoly of the Royal Mint dates back for just over six centuries, but Liverpool and other merchants struck and issued their own coins (for what they were worth) much more recently. And they were worth what they contained in metal, and more when the credit of the issuer was good. These coins enjoyed one peculiarity in which they differed from State coins—immunity from debasement. Edward I. made 240 pennies out of one pound of silver. Edward III. made 270 out of the same weight. Henry IV. increased the number to 360; Edward IV. to 450; and Henry VIII. contrived to make no fewer than 864 pennies out of the like amount. Private coiners never appear to have indulged in these profitable little tricks. They appear to have given full weight and quality, and not to have burdened either their credit or their consciences. It is not so long since an attempt was made to foist a bad half-sovereign upon the people—an attempt by a Chancellor of the Exchequer, who still goes about just as though he had done a very honourable thing. But then, 'the State can do no wrong,' and conduct which would bring a private citizen to the treadmill is a virtue in a State.

And what about the cost of minting? Clearly those who want coins are willing to pay for the manufacture. The coin is more useful than the unmeasured bullion. It cannot be made for absolutely nothing, and those who want medals must pay for them. The gold in the coinage would necessarily be worth a little less to melt down than it would pass for as money. A great deal of stuff is written about minting and issuing. Minting is casting metal in little stamped medals with the guarantor's name on them. Issuing is selling these little medals for what they are worth. The coiner gets his profits for his risk and trouble, and the public gets a manufactured article.

'Yes, this is all very fine, but money must not be treated as though it were a mere question of barter medium. Look at the immense waste of precious metals by wear and tear. A State has credit, and should use that credit to reduce that loss. What should we do if the Bank of England issued five, and ten, and fifty pound coins instead of paper notes?' So it is argued. And then it is pointed out that bone counters, if properly identified and safeguarded against counterfeit, would effect a great saving in precious metals. This looks true, but it shocks the system, and the consequence is that, with his proverbial genius for compromise, the Englishman proposes to jumble the two principles together. 'There is a good deal to be said,' says he, 'for the barter medium principle, and there is a good deal to be said for the credit principle; let us have a bit of both.' So somebody brings in a bad half-sovereign Bill; token money is talked about, and a few persons begin to ask themselves why cardboard half-sovereigns will not suffice just as well as nine-shilling half-sovereigns. The analogy of the bank-note is pointed out, and the metallists scratch their heads. Somebody else thinks it would shake the credit of the State. Another thinks it might lead to an undue issue. What is to prevent the State from issuing an unlimited number of cardboard half-sovereigns or half-crowns? Apparently nothing. The coinage would thereupon be depreciated, and the propertied class would be mulcted. It does not seem to occur to these persons that the token coin would, for precisely the same reason, be depreciated to the exact amount of its deficiency of metal. It would pass for what it was worth, and no amount of hard swearing on the part of the State would make it worth a cent more. The metal in it would pass at its market value, and the promise on its face would pass at the same rate as the notes of the State.

The upshot of this is that the systems of money and of credit are distinct, and should be kept distinct. Money measures man's dishonesty; credit measures his honesty. It is therefore satisfactory to know that over 95 per cent. of the commercial transactions in England are conducted on credit. Credit results in enormous saving, and every encouragement should be given to its extension. All obstacles in the way of, or taxes upon, credit are mischievous. There can be no conceivable reason why banks or private persons should not issue unlimited paper. There is nothing to prevent a pauper in the workhouse from issuing a thousand promissory notes of a hundred thousand pounds each, payable this day twelvemonths. He is permitted, and rightly permitted, to obtain any goods he can in payment for these notes. When the day comes round he fails to meet his liabilities. And what harm is done? How much has he obtained for his promises? Possibly six ounces of tobacco. The public would not absorb paper which did not bear on the face of it evidence of ample security. But this does not militate against the contention that every coin of the realm, whether issued by the State or by private individuals, should be regarded merely as a barter medium, and should, therefore, contain full value within itself.

. We have seen how easy it will be to calculate the value of a given weight of bullion under the new system. The reverse process is nearly as easy, but not quite. A thousand grams of pure gold will be worth a hundred lions. But a hundred lions will not weigh exactly a thousand grams, because they will not only contain a thousand grams of gold, but, in addition, a hundred grams of alloy. Consequently, they will weigh eleven hundred grams. If we wish to know the value of a given number of lions, we must deduct one-eleventh of the actual weight.

Then a lion is not a lion, you say. If the lion of account is ten grams of pure gold, the actually coined lion must be worth at least a trifle more. If the gram of alloy is silver, it will be worth four doits. So that a lion, which is said to have a value of ten grams of gold, turns out to be worth ten grams of gold, plus four doits, plus the cost of manufacture. Anyhow, it is worth more than the theoretical lion. Even if copper is used for the alloy, as it is in this country, it is worth something, and then there is the cost of the mintage. Now it happens that in this country, if you take 113·1 grains of pure gold to the mint, they will put the whole of it into a coin weighing 123·27, and will return it to you without charging you anything for the copper they have added, or for their trouble in making the coin. If you try to be clever and extract the alloy for yourself, and then take the 113 grains of pure gold back to the mint for another sovereign, you will find that the process costs you more than the alloy is worth.

Let us look into this question a little curiously. It certainly seems strange that anyone should give you a manufactured article in exchange for the raw material it contains, without further charge. How does this come about?

You have a quantity of pig-iron, let us suppose; and I have a quantity of gold. We agree to deal. You ask a hundred pounds for your iron; I reply, 'If your iron is just what you represent it to be, it is worth my hundred pounds. But you know very well that it is the custom of the trade that I should come and test your iron, before buying: I must come and break the pigs, and apply other tests, and examine the grain of the iron, and I must see that I get full weight. All this costs me money; therefore, I cannot afford to give you exactly what you ask, but a trifle less.' 'Quite fair,' you rejoin; and the price is fixed. But when I tender you the ninety pounds odd, it is your turn to speak. 'Look here,' you say, 'how do I know that your gold is pure and of full weight? I shall have to assay and to weigh your gold; and all this takes time, and costs me money. I must charge for this: suppose we fix the charge at the same figure you charged me for testing the iron?' This seems satisfactory, and the end of it is that we exchange our wares in the same proportion as though no testing or weighing had had to be done.

But it occurs to me that this is a great nuisance, and that I might find some way of escaping the cost of having my gold assayed. I go to my fellow gold merchants, and I say to them, 'Cannot we form a big company, call it a mint, which will run all our gold into little cakes or coins, and stamp them with a mark or device not easily imitated, signifying that the company guarantees the purity and weight of the coins? You all know very well that it is the vendor who has to pay for sampling and testing. With our reputation for wealth, and for that honesty which springs from good policy, our coins will in time pass muster without further examination, and we shall save this element of expense. Sceptics and rivals will be continually sampling and testing our coins, and when it is found that they are always of full weight and purity they will work their way into general circulation and public confidence.'

This is the first step in the direction of a currency. It is practically what really happened. Originally, no doubt, before minting became a State monopoly, private coiners tried to make a large profit by issuing coins worth less than their face value. This was sure to be instantly discovered, and rival mints were launched on a basis of strict integrity. If the business of coining money had never passed outside the domain of private enterprise, it is quite clear that a small charge would have been made for its manufacture, as it would not pay anyone to make medals for nothing, even if the raw material were freely supplied to them. But free competition would render it difficult for any company to make any considerable profit in the minting business beyond a reasonable return on the necessary outlay and cost of manufacture and the normal rate of profit suited to the character of the business. The mint would be a private concern like the Bank of England, and worked on the same business principles, as Burke said it ought to be.

Having got our mint, or several mints, let us see how they would behave. In the first place, they would not do the gold merchants' work for nothing, and then send the bill to the agricultural labourers and other classes, as the State does now. People bringing gold to the mint to be minted, would be charged a small sum for the job; just as farmers are charged who take their corn to be thrashed elsewhere. It would of course have been discovered, as even our State mint has discovered, that pure gold is too soft, and that it is better to put a little alloy with it in order to harden it. If copper were used, as it now is, the charge for the alloy would be insignificant; still, it would help to swell the total charge to the customers.

Let us now see what will happen to the coins themselves, and for the sake of simplicity let us suppose they are the coins of the basic system. Here is a new lion. It weighs eleven grams; it contains 10 grams of pure gold and 1 gram of alloy. It bears the date of its issue, its name, weight and fineness, together with the device of the company or mint, all stamped upon it. The public are satisfied with it. But the years wear away, and so does the coin. People willingly admit that it is still what it professes to be as to its fineness, but what about its weight? It is found to have lost one-eighth of a gram. The coin is refused, or accepted only for what it is worth. Now this re-introduces all the old bother with the scales. Every coin, after the first year or two of its life, is either put on the scales or refused outright. Under these circumstances, what will the mint do? Remember the principle; it is the *vendor* who pays for all weighing and testing. Consequently, the gold holders go to the mint and ask to be relieved from this new trouble. They are willing to pay a trifle more, if the mint will guarantee the weight of the coins either for ever or for a fixed period. A very small premium indeed will enable the mint to do either one or the other. The premium would go to a sinking fund to defray the expense of giving new coins for old, either at any time, or during a period of, say, ten years. Coins dated 1894 would pass by tale up to the end of the year 1904, and after that date by weight only. It is hardly probable that a private company would insure the weight of their coins for an unlimited period, as our Chancellors of the Exchequer seem to have thought it reasonable for the State to do. Without for the present maintaining with Burke that the manufacture of coins should be left entirely to private enterprise, it is allowable to contend that if the State persists in engaging in coin manufacture it should do so in an economical and business-like way. Instead of this, it not only flatly refuses to manufacture coins out of any other metal than gold, but it also forcibly prevents private persons from doing so. It is hardly necessary to point out that there is no silver money in this country. Of silver tokens there are enough and to spare. But tokens will never do the work of real money. Our English shilling is about six-pennyworth of silver, bearing the State's promise to pay another sixpence on demand. What precisely the life of our English silver token coins is I do not know. Some authorities suppose it to be fifteen years, others seventeen, and others less. Let us suppose it to be ten years. Then, on the average, there will always be about ten years' minting in circulation. In the last ten years we have coined about ten million pounds' worth of silver coins, face value. Therefore, there will be about ten million pounds' worth in circulation; of which five millions or thereabouts is money, and the other five millions is the State promise to pay on demand. This means a debt of five million pounds, which ought to be added to the national debt. Or rather, it has been added to our indebtedness, but not reckoned in the debt. It is practically a loan upon which the State pays no interest whatever. Thus the Exchequer appears to gain about a hundred and fifty thousand pounds a year on its token silver. But if, by any chance, all our silver coins were presented for payment at once, it would cost, having regard to wear and tear, over five million pounds to redeem them. Similarly, if the public made up its mind to use honest shillings and sixpences, or other silver coins, instead of half-tokens, it would cost five million pounds to rehabilitate our silver currency. The effect of the change upon the Exchequer would be in *appearance* this very heavy loss, but in *fact* it would only be the payment of a debt for which it is ultimately liable. There is a very strong feeling, and a growing feeling, in the country that there is not sufficient gold in the world for a universal currency. Indeed, this is not difficult to prove. Nearly all civilised nations are, at the present time, money starved; the effect of the introduction of a true silver currency concurrent with the gold one would probably be in the highest degree beneficial to trade and commerce. But this would be so only on condition that the step was based on true economic principles, and not upon those advocated by the bi-metallists. Anything of the nature of a national or an international guaranteed ratio between silver and gold would be in the highest degree foolish, mischievous, and dangerous. But, properly effected, the introduction of a silver currency would produce several desirable results.

The first effect of the change worth notice would be a slight rise in the value of silver, brought about by the increased demand; a rise not in itself undesirable, if effected by natural and honest means. In all probability, the rise would only be slight, for the unprecedented issue of nearly four million pounds' worth of silver in 1889 and 1890 had no very marked effect on the market price of silver—though this was, it may be supposed, the main object of the unusual issue.

Again, the proposed change would be a first step in the direction of an international currency, which must be based on fact, and not on faith. A prophet hath no honour in his own country, but a token coin hath no honour out of it. An English pound sterling, like Wellington's soldiers, will go anywhere and do anything. A coin which is what it professes to be is readily accepted by foreigners and everybody else. It can be taken to the nearest metal-merchant's and sold for its face value, or very nearly so. Its value is not dependent upon the credit of the country of its issue. That is the secret of the success of our English sovereign. It is no liar, the world has found that out. But the poor shilling, what can it do? The further it travels from the shores of England, the more contemptible it becomes. Even the credit of rich England will hardly carry it across the Channel. I do not say it sinks to bullion value quite so near home as Paris. But virtue passes out of it. If you go to a theatre or a racecourse with your pocket full of shillings, the best you can do with them, even in Paris, is to get them accepted for francs. Go to Siam or Siberia, and you get the market value of the silver that is in them, and that is about all—fivepence each. Well, that is not the coin for international currency. We may believe that our Government will pay up when its token coins are presented for payment, but foreigners are more or less sceptical, and besides, they have not our experience and knowledge of the coin itself. It may or may not be a shilling. The only certain test of its value in that case is its weight and purity. If that is all right, the foreigner does not care two straws whether it is worth sixpence or twelvepence or twentypence in England; all he knows about it or cares about it is that it is what it pretends to be—a certain weight of good silver. We are always finding fault with the stupid Hindoo for keeping the rupee up at a fancy price in the interior. But how in the world is he to learn that the rupee is not intrinsically worth what it was, when every English coin he see lies to the contrary? He may be a fool at political econony, but he is wise enough to know that his two-shilling piece is worth more than ours, anyhow. Take an English florin and an Indian rupee to a Chinese silversmith, and he will give you more for the rupee than he will for the florin. Take the same coins to a Calcutta banker, who has faith in English Government credit, and he will give you more for the florin than for the rupee. What we want is a coin that will fetch the same price at the Chinese silversmith's and the Calcutta banker's. That is an international coin.

Another curious, but not very important, effect of the establishment of a true silver currency will be the immediate effect of the change on prices. There can be little doubt that it will at first tend to lower the prices of all commodities usually sold for silver; that is to say, to lower retail prices. Such is the simplicity of human nature, that there will be a demand for the new coins above their intrinsic purchasing power. Poor persons will not readily give a coin which looks like an old-fashioned half-crown for something which used to cost one shilling and fourpence halfpenny. And even the retail shopkeepers will at first be willing to take something less.

A good deal of fuss is made about the wear and tear of the coinage, and the actual loss of gold and silver consequent thereon. And advocates of our present silver token coins point to the saving in wear and tear due to the fact that such coins only contain about half the quantity of silver which real coins would contain. But, to tell the truth, the cry is mostly got up by those who have their own axes to grind, and who have an interest in the debasement of the silver. Nobody pretends that the loss of good metal by friction is a good thing in itself. But everything which is useful has a tendency to wear out with use; and there is no particular call for a special providence to make an exception in favour of coins. People who cannot trust one another cannot deal wholly on credit. Therefore they must use money of some kind. Money wears out and has to be replaced, and those who need it must pay for it. Coins form no exception to a universal rule. Credit is, doubtless, better than money when people are honest, either from virtue or policy. As folks improve, credit increases. But it cannot be artificially increased. It is a natural growth. The stronger the general credit in a country, the less the money required to carry on the business of that country. It is said that England makes 20 shillings' worth of coin do as much trade as 85 in France, 40 in Italy, 58 in the United States, and 100 in Spain. It is simply false to pretend that the public likes mean, attenuated coins. It is not so. People prefer honest, plump, full-bodied coins. The reason why there are so few crowns and four-shilling pieces around is not so much because those coins are too heavy, as because there are so few commodities of which they represent the exact value. When the old copper penny was called in, there was much gnashing of teeth at the mean-looking thing that took its place.

It is needless to say that though the public, and more especially foreigners, might be willing to trust the mint company as to the weight and fineness of its coins, it by no means follows that they would be willing to pin their faith to its notes, or, what comes to the same thing, to its tokens. They might or they might not. In other words, the silver coins issued by a private mint would have to be worth their face value. The mint would not be able to issue five-pennyworth of silver as a shilling. I am not saying a word against token money as such. I believe the free issue of notes, and therefore of tokens, when based on natural credit and regarded simply as symbols of credit, to be a good thing. Trade on a credit basis is more economical than trade on a money basis. Money measures man's dishonesty; and dishonesty is costly. This is altogether a separate question.

What is now the position of our silver currency? The market ratio of silver to gold is at the time of writing 34. Therefore a shilling ought to weigh about $\frac{17}{16}$ of a sovereign. In other words, a shilling should contain by weight $1\frac{7}{16}$ as much pure silver as a sovereign contains pure gold. But a sovereign contains 113·1 grains of gold; therefore, a shilling should contain 192·25 grains of silver. What does it contain? 80·7 grains; or considerably less than half its face value. It follows that at an outlay of about £80,000 the State can at any time put into circulation no less than £190,000 worth of silver, face value. Chancellors of the Exchequer have been very busy with the silver currency of late. It is true the Exchequer does not pocket the extra £110,000 thus created, as it were, out of nothing; because the silver finds its way back to the mint in time, and it then has to be redeemed at its face value. But what it does pocket is the interest on the £110,000 during the average life of silver coins. And that life is somewhere between ten and twenty years.

Five-twelfths of our silver currency, then, is money and the remaining seven-twelfths credit-notes. So that our State with its mint is virtually a big bank; and, what is more, it is a bank with a very large issue of notes resting entirely on its credit, and without even a pretence of a gold reserve to meet its liabilities.

Clearly, anyone has a right to get his silver converted into coins, or into any other manufactured article. But he has no right to claim a State guarantee or any other guarantee that his coins shall be accepted in exchange for a fixed quantity of gold; because this is tantamount to compelling the State or someone to buy silver when they do not want it.

Let us now suppose that there are several private mints; it is obviously the interest of all such mints to coin any amount of silver brought to them, and, if required, to guarantee such coins for a limited period. And their rates might differ—probably would differ slightly—but no one of them could pledge the others to redeem its issue. Hence there might be some half-dozen issues of half-crowns and sixpences bearing the names of different guarantors, but otherwise alike. This would be extremely confusing and inconvenient, and the result would be either that the public would gradually refuse all but one brand, say that of the Bank of England or of some union of banks, or find some other way out of the difficulty. The way out is not very hard to find. Uniform coins are required. Then let them be uniform. Let all the crowns be alike or all the crosses; let all the shillings be alike or all the half-crosses; according as we use our present coinage or that of the basic system. And let the names of the guarantors of the several issues be kept in the mint books. The public would then look to the State, as we do now. But this is very different from the present arrangement; for the State would look to the guarantors and would run no risk, and the enormous advantages of free competition would be secured. In no other way can the right and true premium be ascertained. What that is, it would be presumptuous in me to pretend even to guess.

What is the natural process whereby anyone should be able to obtain coins of guaranteed ratio in exchange for his gold or silver bullion? The thing is very simple when not obscured by currency theories.

A gold-merchant comes to the mint with 10,000 grams of gold. He wishes to have it coined into 1,000 lions. 'All right,' says the mint master, 'we can do that for you. This is what you ask us to do :—

'1. To assay or analyse your bullion.
'2. To separate or extricate the impurities.
'3. To add the necessary copper or other alloy.
'4. To mint the coins.
'5. To guarantee the weight and fineness for 10 years.
'6. To give you in exchange for any one of them at any time 10 silver crosses.

'With one exception, all these elements of cost are pretty constant. We know what they amount to. It is 4 doits per coin for the first four processes; and then there is the fifth. Now we estimate that each lion loses per annum by wear about ·0005 of its weight. In 10 years, therefore, it will lose about ·005. This is, of course, ·05 of a gram, or 5 doits. Allowing for discount, and for coins which never return to the mint at all (*e.g.* those lost at sea) and for those which return before their time, we can afford to charge you 3 doits. Then there is the sixth item. No rise in the value of silver seems to be generally expected. Consequently the London and Westminster Bank, which has sent in the lowest tender, offers to cover this risk for ·2 per cent., which makes another 2 doits. This makes our total charge 9 doits per lion, which is exactly $L9$ for your thousand coins. That is our charge. Or we can deduct it from the total, and hand you the balance of 991 lions.' It follows that a lion piece is really worth more than a lion of account. And the jeweller who melts down coin of the realm is a fool for his pains, for he could have got his thousand grams of gold bullion for 991 lions.

Similarly, a silver-merchant comes along. He has 25,000 grams of silver, which he wishes to have minted into 1,000 crosses. 'All right,' says the mint master, 'here are our charges: for assaying, refining, alloying, and minting, 4 doits per coin; for guaranteeing weight and fineness for 10 years, a half-doit. Then you ask us to do something more for you than this: you ask us to guarantee the exchange of your silver coins at a stipulated ratio for gold—in other words, not only to give you a cross containing 25 grams of silver, but also at any time within 10 years to give you one-tenth of a lion for it, no matter what change in the market price of silver may have taken place in the interim. This is a very different matter. If silver goes up we shall never see these coins again, but if it goes down they are sure to be presented for redemption. At present the price of silver fluctuates considerably, and in the main it seems to be steadily falling. In order to insure you against loss, we find we shall have to charge you $5\frac{1}{2}$ doits. This brings our total charge up to 10 doits per cross, which is 10 per cent. of the silver to be coined. We can, therefore, hand you at once 900 crosses for your silver bullion.'

I have no knowledge, nor has anyone else any knowledge, of the precise charge which would be made for insuring the ratio of silver to gold for a fixed term, because the business has never been thrown open to free competition. It would certainly amount to a very considerable fraction of the value of the silver coinage. Nor is there any other way of ascertaining what the charge ought to be.

If this were done, I see no formidable objection to making silver legal tender to any amount in the absence of a special clause to the contrary in contracts of sale. 'If I understand the matter aright,' says Jevons (*Money*, page 77), 'every person is at liberty to buy, sell, or exchange in terms of any money or commodity whatsoever which he prefers. It remains quite open to a creditor to receive payment in coins which are not legal tender if he likes to do so, and I presume there will be nothing to prevent him entering into a contract to that effect.' In other words, legal tender is simply a State definition. In the absence of qualifying words, the State gives notice that by the words 'a hundred pounds' or 'a hundred lions' it will understand one hundred English golden sovereigns or one hundred golden lion pieces. Clearly, then, there would be no State interference with the freedom of the subject if the State were to give notice that in future it would understand by $L100$ either a hundred golden lions or a thousand silver crosses. Those traders who preferred to deal in gold could say so; those who preferred to deal in silver could say so; whilst those who did not care a straw in which they bargained need say nothing. The case of the bi-metallist is based on the assumption that the majority of traders would actually prefer to deal on the optional system, because, so it is alleged, there would be less likelihood of a simultaneous rise or fall in the value of both metals than in that of either one of them. If this is so, there seems to be no valid reason why the plan should not be tried, and those who do not like it can contract out of it. But in either event the debtor would gain. In order to avoid this, let the State insure both the silver and the gold coin. We have seen what the bankers would say about insuring the silver coins. What the bankers *would* say the State *ought* to say, if it undertakes the function at all.

Once more, what is it which the State is asked to do? We will assume, in face of the fall in the price of silver, that the bi-metallists have thrown overboard the absurd ratio of $15\frac{1}{2}$, seeing that the present ratio is about 34—though the ratio in no way affects the principle. The State is asked to do for nothing what, as we have seen, the bankers would probably charge something like 10 per cent. for doing—namely, to guarantee the holder of silver coins that at the end of 10 years, or perhaps for an indefinite period, he shall be able to come to the State and demand $\frac{1}{25}$, or whatever the chosen ratio may be, of the weight of his silver in gold. The State is also asked to guarantee the holder of gold coins that at any time he may be able to bring his gold coins to the mint and demand twenty-five times their weight in silver. The bankers would of course be willing to do this at a reasonable premium. What that premium is nobody knows. Of course, those who insured equal quantities, measured in value, of both metals would not lose on both transactions if the premium was properly calculated in the first instance. If silver went up, they would have to pay the gold holders; if silver went down, they would have to pay the silver holders; and the premium they receive for the one would go towards the redemption of the other, and would on the average cover it and leave a small working profit to the guarantors. But is the State at all likely to calculate the chances correctly? And if not, why should the tax-payer be compelled to undertake such a risky function? If bankers or financiers or any union of them or any special coin-insurance company care to take the business up, there can be no objection to their doing so. Similar speculations are of every-day occurrence in dealing with other commodities. And there is nothing whatever to urge against putting gold and silver on the same footing as iron and coal, provided it is done through the channels of private enterprise and free competition, and not by State agency. The only obstacle in the way of this obvious and rational process of establishing a double currency is the still surviving superstition, shared by all parties in the financial world, that gold and silver are *sui generis*, and not subject to the same economic laws as other valuables.

I have said that at the time of writing the ratio between silver and gold is about 34, but I have little doubt that the revived demand for a true silver currency would speedily bring it down to about 25, and I have based my proposed scheme of coinage upon that assumption. At the same time it must not be supposed that the principles of the system are in any way affected by the exact ratio selected at first or ascertained at last. Bi-metallists, at all events, are barred by their own utterances from contending that the ratio will deviate much from that which is finally agreed upon between the State and the guarantors, after a few years' experience.

'What is a pound?' asked Sir Robert Peel, and from that day to this the question has remained practically unanswered. A pound is a certain weight of pure gold, namely 113·1 grains; or it is 123·27 grains of standard gold—that is to say, of gold $\frac{11}{12}$ fine. If it is one it cannot be the other, and yet as often as not the same writer will treat it now as one and now as the other. An ounce Troy of pure gold is worth 4*l.* 5*s.* all but a halfpenny, and an ounce Troy of standard gold is worth 3*l.* 17*s.* 10¼*d.* Does a promise to pay ten pounds mean a promise to pay 1,131 grains of gold? No, it means 1,131 grains of *proved* gold, weighed and assayed, and it costs the debtor about ten pounds and fourpence to get it; and the system adopted in civilised countries is that of getting the gold guaranteed as to weight and fineness by the State. Why not treat silver in precisely the same way? Bear in mind that the State does not guarantee the value of gold in terms of any other commodity, or in any way; it merely guarantees the fineness of its own coins, and their weight for a defined, or in some cases for an undefined, period. Why should it not do the same for silver, or, if need be, for tin or lead? There is no reason whatever to the contrary. But mark the consequences. Remember, we have no silver coin in this country. A gold coin is not a promise to pay. A bone token is a promise to pay, and only that. A shilling is partly one and partly the other. It may be called a token coin. Moreover, our silver token coins do not consist of a promise to pay a certain fixed weight of silver, but of a promise to pay so much gold. A crown is a promise to pay 28·3 grains of gold in exchange for it, and its intrinsic value as a bit of silver is now 11·35 grains of pure gold, and it is therefore worth about two shillings and a penny. What is wanted by the advocates of a double currency is a silver coinage treated in all respects in the same way as our gold coinage. It is said, with perfect truth, that there is not enough gold above ground to serve for a universal currency, and that therefore it should be supplemented by silver.

But now suppose we had true silver coins weighing, say, 25 grams each, and purporting to be simply that and nothing more; not professing to represent a fixed weight of gold, as our present token coins do. What would the consequence be? Nothing would prevent slight variations in the relative value of gold and silver; hence we could not go into the market with gold and silver coins in our pocket, and rely upon receiving the same change from day to day. One day a pound would be worth twenty shillings and twopence; another day it would be worth nineteen shillings and ninepence; and so forth. The effect would be confusing, and it would lead to endless disputes. At the same time, such silver coins would travel all over the world just as our sovereigns do now. In order to make them useful and convenient for ordinary transactions at home, it would be absolutely necessary that there should be some place where the holders of them could get them exchanged for a fixed proportion of gold coins. The demand for this convenience would cause persons to spring up spontaneously to meet the demand, and unless they were prevented by the State, certain money-changers and financiers would be certain to put their own stamp on the silver coins and to issue them at a guaranteed ratio for a small premium. Not only would this disfigure the coins, but it would also lead to fraud, and, moreover, it would not meet all the requirements of the case. What objection is there to having this done by the State? There is this objection: a premium would have to be charged on each minting of guaranteed silver coin, and this premium would vary from issue to issue. Without competition—and the care, thought, and experience which competition always entails—the premium would be either needlessly high, which would be a burden on customers and a tax on currency, or too low, in which case the taxpayer would be plundered.

If the bi-metallists are right, those bankers among them who believe that the price of silver would rise on the adoption of a silver currency would guarantee the ratio for ten years for next to nothing; provided such guaranteed ratio started at the then market price. In order to make the scheme perfect and consistent, it will be desirable to have the gold coins also guaranteed in terms of silver for a like period of, say, ten years from the date engraved on the coin. The objections which would probably be raised to this reform are, first, the expense of calling in the whole of our silver coins, the cost of which, including their redemption, would amount to over five million pounds sterling. But the major part of this would be merely the repayment of the debt for which the nation is eventually liable; and as for the balance, it would probably turn out to be a splendid and profitable investment. Secondly, it has been suggested that the banks would form a ring to keep the premium extortionately high. I think there will be little fear of this, and certainly such a ring would speedily be broken. It is also urged that the transaction would be of too speculative a nature to attract high-class banks; but in these days of marine insurance, of life insurance, of accident insurance, and the like, we may rest assured that there will never be any difficulty in finding persons of stability to undertake any reasonable risk at a premium ascertainable by competition. It has been urged that out-of-date coins would continue to circulate among unwary and illiterate persons, especially in the country, but the probability is that few coins would live out their allotted time, and if the worst came to the worst the last holder would lose only the difference between the face value of the coins in his possession and their value as bullion.

The strongest objection of all is undoubtedly this, that it would probably be necessary from time to time to alter the ratio, in order to prevent the premium from getting altogether too high, and so reducing the coin to the condition of a token coin. On these occasions we should have two sets of coins having the same values but different weights. The answer to this is, and bi-metallists at all events cannot deny it, that in all probability the ratio when once ascertained would remain fairly steady, and the fluctuations would be so small as not to call for any alteration in the ratio. But even if this were not so, no great harm would result. The light coins would remain in England and speedily find their way to the mint, while the heavy ones would travel and live out their appointed time. There remains the sentimental objection to the disappearance of the venerable pound sterling, the 'fixed star' of our system; but we cannot allow too much weight to mere sentiment. It is further said that the change would occasion great confusion in regard to existing contracts. No such confusion could arise in large transactions, because the ratio of the pound to the lion would be well known to be £1 to £1 7s. 3½d. And the ratio of the lion to the pound is $L1$ to $L·7988$, a ratio well within a hundredth part of a farthing.

The system of coining gold in this country is neither very well understood nor indeed quite intelligible. Any person may demand of the Issue Department of the Bank of England notes in exchange for gold bullion at the rate of £3 17s. 9d. per ounce of standard gold, to be melted and assayed by persons appointed by the bank. But since the true value of the standard gold is £3 17s. 10½d., it would seem that the person who tenders the bullion loses three-halfpence per ounce. It is said by some that this three-halfpence goes towards the expense of the minting; but this can hardly be the case. For, as Mill points out, 'though there is no seigniorage on gold coin (the mint returning in coin the same weight of pure metal which it receives in bullion), there is a delay of a few weeks, which to the holder is equivalent to a trifling seigniorage;' that is to say, the mint pays three-halfpence an ounce more for the gold than the Bank of England. In fact, it pays the full value. But although it matters very little to the person tendering the bullion whether he forfeits three-halfpence at once or is kept out of his money for a few weeks, it cannot be said that the mint gains the three-halfpence, whatever the bank may do. For the nation would be none the richer if it stacked the gold for three years instead of three weeks before minting it. The loss to the holder is no gain to the mint. Consequently, we are justified in saying that the Government in this country coins gold gratis for anyone who furnishes the metal. And since gold holders usually go to the bank rather than to the mint, it follows that the mint actually pays about ten pounds and fourpence for every ten pounds' worth of gold that it purchases. This is a mistaken policy, and one of the consequences is that some foreign nations, such as Portugal and Brazil, save themselves the expense of minting gold by the simple process of circulating our English sovereigns and half-sovereigns, and they also reap the advantage of our gratuitous guarantee against loss by weight, to the extent of about three-quarters of a grain to the sovereign, and customarily even more. All this would be avoided, together with other inconveniences, if the customer bringing gold to the mint were charged the full cost of assaying and minting and of indemnifying against wear for a fixed period, and at once received gold coins or notes for his bullion without further delay. The existing arrangement merely obscures the fact that a sovereign is really worth more than the 113·1 grains of pure gold: just as any other manufactured article is really worth more than the raw material of which it is made. I have laid stress upon this somewhat trifling fact because, although it is of little practical importance when we are dealing with gold coins, it assumes very different proportions when we come to deal with silver coins, and still more formidable proportions when we come to deal with a double currency, in which the ratio of exchange between the gold and silver coins is guaranteed. It is the failure to recognise this which has led our bi-metallists astray. Supposing only the silver coins to be guaranteed in terms of gold, it is said that silver coins thus guaranteed would, if the value of silver rose, speedily fall into the hands of the silver merchants. And if, on the contrary, it fell, our gold coins would find their way to the melting pot. For clearly the silver coins would be still worth their face value if the price of silver fell, and they would be worth more than their face value if the price of silver rose. Hence they would be preferred and held back, and, according to Gresham's law, the gold coins would circulate and we should have a purely gold currency. If, on the other hand, the price of silver fell, the gold would be withdrawn from circulation, and we should have a purely silver currency. This argument overlooks the fact that a silver coin would in the nature of things be worth at least ten per cent. more as a manufactured article than the pure silver it contains. What the bi-metallists are asking the State to do is to give them an article worth twenty-two shillings or more for a sovereign. This is a request which a wise nation is hardly likely to grant. The difference between a sovereign and a pound sterling of account is precisely the same in kind (though not in degree) as the difference between a ton of coals at the pit-brow in

Yorkshire and the same ton of coals on a London wharf: it is the same thing, only more conveniently adapted for use. The minting of the gold and the carrying of the coal, without altering the intrinsic nature of the commodity, have made it a more convenient article.

It is said, and truly, that our present system compels us to speculate in gold. True; every contract made in terms of pounds, shillings, and pence is necessarily a speculation in gold. During the depreciation following on the great gold discoveries, creditors lost millions. Since then, during the recent appreciation of gold, debtors have lost heavily. Prices are now so low that annuitants, pension holders, and rent receivers in England can buy more with their fixed incomes than they formerly could. All money transactions between gold-using and silver-using countries have been seriously disturbed, and are always in danger of being further disturbed, especially those money transactions which extend over long periods of time. But is there anything in the law to prevent people (say Lancashire and India merchants and manufacturers) from making their contracts in ounces of silver or tons of copper or tin or iron if they choose? No one is compelled to buy gold, any more than he is compelled to buy wheat. It is usual, but not obligatory, and if traders mutually agree to deal in silver they can do so, and no legislation is required for the purpose. If A. sells a cab-horse to B. for 64 ounces of silver, he cannot claim to be paid in gold. The price to-day would be 8*l*., but if it fell to 7*l*. A. would have no redress, and if it rose to 9*l*. B. would have to pay. Why, then, do people not deal in silver? The answer is simple: because there are no silver coins. It is the old story—they do not want the trouble and expense of weighing and assaying the metal at every deal. If the mint would coin ounces of silver or twenty-five gram silver coins, with or without alloy, people would deal in silver. Or if the State would get out of the way, private mints would coin silver. And silver coins would have as universal a currency as gold coins. Mr. Childers's lamented gold nine-shilling half-sovereigns would hardly have got across the Channel.

I have said that in order to render a double currency complete and perfect it would be necessary not only to guarantee the silver coins in terms of gold, but also to guarantee the gold coins in terms of silver. Eventually this would probably be the best course, but as a first step, and as a transitional measure, it might be wiser to guarantee the silver coins only. There is no doubt that the more consistent and complete scheme of mutual guarantee would increase the cost of the gold currency, and advocates of the single gold standard would perhaps have a right to grumble at this. But they would have no right to complain at the introduction of a concurrent silver currency guaranteed in terms of gold, because they would be under no compulsion to make any use of it. If there is a respectable section of the community anxious to have real silver coins, and willing to pay for them, it is nothing less than despotism to refuse it. Personally I have little doubt that, within a few years of its introduction, the gold-holders would themselves clamour for a gold currency guaranteed in terms of silver. Of this at least there can be no reasonable doubt, the double guarantee of the two metals would greatly steady the value of both.

SPECIMEN OF ACCOUNT BOOK

Date			Crosses	
Feb. 23	6 y. sirloin beef at 6 G. per y. .	. .	3	6
,,	200 eggs at three a G.	. . .	6	66
Feb. 26	1 peck of potatoes .	. . .		35
,,	Dozen Champagne (Koch & Fils)	. .	40	
Feb. 27	5 sacks kitchen coal .	. . .	3	75
Mch. 3	Gross of Swedish matches .	. .		25
,,	Box of figs	. . . .		3
Mch. 12	2 tons best Silkstone .	. . .	15	75
,,	1 pot currant jelly	. . . .		15
Apl. 4	3 l. olive oil	. . . .	2	5
,,	Stamp	. . . . .		02
		Total .	. 78	38

It is clear that such an account as this would not require a number of ruled columns like those now in use. One 'thin red line' would suffice. The figures entered on the left-hand side of the line would denote whatever unit was most suited to the account. Thus in large dealings, as between a broker and his client, the line would be drawn between the lions and the crosses; in small retail transactions, it would be drawn between the crosses and the groats; and in very humble tally shops, it would be drawn between the groats and the doits. The unit employed would be stated at the head of the column, as shown above.

Consider the simplicity of the addition. Here is no turning of farthings into pence, pence into shillings, and shillings into pounds. There is nothing but simple addition. The total may be read as 7,838 doits, or as 783 groats and 3 doits, or as 73 crosses, 3 groats, and 3 doits, or as 7 lions, 3 crosses, 3 groats, and 3 doits; and in any case it would be written simply 7,838.

Again, consider the simplicity of calculating the prices of any given quantities. This sugar is 25 (that is, 2 groats 5 doits) the yasp. I want a ton of it. I simply add three naughts, and the thing is done. The price is 25,000, or 25 lions. Similarly, if I want a mere lump of sugar to sweeten a pie, say 10 grams, I simply take off two naughts, and I see that the price is 0·25, or a quarter of a doit.

There would of course be no objection, for the purpose of abbreviation, to the use of symbols like $l.$ to denote the particular denomination of coin under consideration, in books and papers, though they would not be required in accounts. Thus 7,000 might with advantage be written 7$l.$ as now. And, what is more, in dealing with large amounts like the National Debt, or Imports and Exports, it would be simple to write in terms of tons.

CHAPTER XV

SUMMARY

BEFORE taking leave of the subject, it may be well to review the several tasks I have endeavoured to accomplish.

I have tried to show that the original Aryan system of measures was based upon the yard; that the unit of bulk measure was a cylinder one yard high and one yard in diameter, variously called *culeus* or pipe, or by any other name.

Halving the dimensions, we have another cylinder one-eighth of the bulk, and by continuously repeating the process, we find ourselves in possession of what I call the pipe scale. All these measures of capacity are represented by modern English measures about one thirty-second smaller than the originals.

I have followed the pipe scale back to Roman and Greek times, and shown, or endeavoured to show, that the bulk-measures of both these peoples were identical with our own. I have explained the fallacy upon which is based the otherwise absurd theory that the Greek bushel was out of harmony with the Greek peck and the other popular measures. In this at least I think I may claim to have succeeded. I have dragged to light for the first time the plan upon which the Roman classical system was based, and the ingenious scheme by which its founder minimised the practical difficulties attendant on the introduction of all new and unfamiliar arrangements. The new *quadrantal* was found to be so related to the old *amphora* that the former contained three *modii* within an extremely small fraction, while the latter contained four *modii*. The *modius* and *sextarius* (peck and pint) would remain practically what they had always been under the old binary *régime*.

I then set out to show—what has often been suggested—that the standard weights of the pipe system were co-ordinated with the measures of bulk through the medium of wheat; and I think my readers will agree that I have demonstrated that the specific gravity of the wheat used by our ancestors was ·8, or sixty-four pounds to the bushel, and not sixty-two pounds to the bushel, as is usually supposed.

Having now restored the Aryan system of weights and measures, I applied the results of our investigation to the tangled skein called the Greek system. I inferred that, having a series of bulk-measures based on their own yard or half-yard (it matters not which), they would probably adopt the same principle for establishing a weight series. But it became increasingly and indeed overpoweringly manifest that the specific gravity of the standard wheat was *not* the same as the Gothic, but that the wheat used by the Greeks as the medium was five-sixths of the weight of an equal volume of water. It then appeared that the Euboic talent, which probably had nothing to do with Eubœa, was exactly five-sixths of the Æginetan talent, and the natural conclusion almost springs to one's eyes that the latter was the weight of some measure of water while the former was the weight of the same measure of standard wheat. What was that measure? I think I have shown beyond a shadow of a doubt that that measure was the ἀμφορεύς—not the ridiculous ἀμφορεύς of our school tables, but the real Greek bushel.

Before referring to the Solonian talent I must first point out the extraordinary coincidence which attracted my attention while trying to discover why the Saxon pound should have taken firm roothold in a country using the beautiful and simple pipe scale, with its pound or pynd the weight of a pint of wheat. This Saxon or Colonia pound seemed so meaningless and anomalous that I could not account for its survival and even popularity, till it occurred to me to ascertain whether it was after all the weight of a popular measure of any other substance. It will readily be imagined that I was agreeably surprised when I discovered that the half-nail cylinder of pure gold weighs exactly a Colonia pound—not within an ounce, not within a pennyweight, but to the *tenth part of a grain*. Here was food for reflection. Taking the well-known ratio of the Solonian talent to the other two talents, I adopted the same method, and was pleased (but not surprised this time) to find that a Solonian talent was the exact weight of a χοινιξ of pure gold—a Greek quart-pot of gold. The expression used by Herodotus and others, ταλαντον ἀργυριου, in reference to the Solonian talent, is now as clear as noonday. For I have explained how the gold and silver weights were co-ordinated both by Solon and by the Roman moneyers.

I should, however, have been disappointed in this quest had it not been that I had previously made a necessary correction in the weight of the Colonia pound. Hitherto it has been set down at 5,400 Troy grains. I have shown that it weighed a pennyweight more, or between 5,420 and 5,421 Troy grains; and in demonstrating this I have detected the moment and explained the cause of the birth of our modern Troy pound.

I have given in my adhesion to the theory that the yard itself had a scientific origin of the highest order, being the double of the Egyptian cubit, which was the thousandth part of the mean distance travelled by the earth's surface in one second of time at the equator. Our own yard is still within a very minute fraction of this double-cubit.

Following the old pipe system down from the earliest times to the present day, we note several shocks or cataclysms. The first may be called the Sillian metric revolution, when the duodecimal scale took the place of the binary, and bulk measures became cubical instead of cylindrical.

Again, when the Colonia pound was brought to this country by the Saxons, the grain was increased by one-fifteenth, while the ounce remained the same, and the pound took the place of the pynd for money weight.

Then, after numerous currency juggles and much talk of true

weight or treu weight, we come down to 1527, when Henry VIII. (*perhaps* inadvertently) invented Troy weight, and by so doing contrived to make a penny in every pound for himself. From that time the old pipe system crumbled more and more into decay, and received its *coup de grâce* in 1789, when the metric system saw the light.

The question next presents itself, Shall we make a clean sweep of the *débris* of the old system and embrace the metric, or shall we restore the old pipe scale? There is a good deal to be urged in favour of both proposals. I am convinced, however, that the balance of advantage is immensely in favour of the metric system, provided we observe the rules based on the experience of ages, and above all reject and destroy the French nomenclature, which has proved so disastrous and retarded the universal adoption of the system for a century.

In Chapter XII. I have examined the requirements of a really good system of measures. Guided by the principles evolved from a survey of several conflicting schemes—especially the old pipe scale, the Roman cubical system with duodecimal subdivisions, and the French cubical system with decimal subdivisions—I have constructed a basic system which is simply the French measures without the French nomenclature, and with a superior scheme of co-ordination. To some (not the most thoughtful) it may seem that the names I have conferred on these measures are fanciful, and that the plan on which they have been selected is verging on the childish. But those who are acquainted with chemistry and with the enormous benefits conferred upon it by its admirable nomenclature will need little persuasion to accept some such measure-names as those which I have, after careful study, chosen.

In selecting the unit of value it is important to avoid adopting an artificial and unnatural medium such as an alloy or a metal so many parts fine. The medium chosen should be *pure*, and the coins themselves can then be alloyed to any extent that seems desirable in order to make them harder, prettier, more lasting, or less easily imitated. The French made the mistake of taking a monetary unit only nine-

ARYAN PIPE SCALE

Diameter	Name (pipe scale)	Bulk	E. liquid	Old Roman	Greek	Gothic grains
Yard (διπηχυς)	Double-pipe Pipe Half-pipe	— — Quarter	Tun Butt Hogshead	— Culeus —	— — —	— — —
Cubit (πηχυς)	Double-boll Boll Half-boll	Coombe Strike Bushel	Barrel Kilderkin Firkin	— — Amphora	— — 'Αμφορευς	— — —
Span (σπιθαμη)	Double-peck Peck Half-peck	Tod Stone Clove	— — Gallon	Urna Modius Semimodius	— 'Εκτευς 'Ημιεκτον	— — —
Hand (palmus)	Double-kan Kan Half-kan	Forpit — Pynd	Stoup Quart Pint	— Congius Sextarius	Χους Χοινιξ Ξεστης	32,768 16,384 8,192
Nail	Double-gill Gill Half-gill	Mark — —	— Gill —	Cotyla Quartarius Acetabulum	Κοτυλη Τεταρτον 'Οξυβαφον	4,096 2,048 1,024
Half-nail (old pollex)[1]	Double-ligule Ligule Half-ligule	Ounce — Skilling	— — —	Cyathus Ligula —	Κυαθος Μυστρον Κογχη	512 256 128
Quarter-nail	Double-dram Dram Half-dram	— Penny Sceatta	Fl. dram — —	— — —	Χημη Κοχλιαριον —	64 32 16
Eighth-nail	Double-carat Carat Half-carat	— Styca —	— — —	— — —	— — —	8 4 2
—	—	Grain	Drop	—	—	1

[1] When the uncial system came in, the pollex, which was originally $\frac{1}{16}$th of a cubit, was degraded to $\frac{1}{12}$th, so as to make it the $\frac{1}{12}$th of a foot.

But, if our metrologists are right, a worse fate befell the Greek δακτυλος, which is now *regarded* as the $\frac{1}{16}$th of a πους, instead of $\frac{1}{12}$th of a πηχυς.

There is much dispute and uncertainty about the δοχμη, διχας, δωρον, and other small length measures; and we may believe that a people which halved the οργυια for the διπηχυς, and that for the πηχυς, and that for the σπιθαμη, would also halve and quarter the σπιθαμη.

tenths pure, and consequently a thousand francs is not 5,000 grams of silver, but 4,500 grams of silver and 500 grams of alloy.

In Chapter XIII. I showed the simple application of my proposed nomenclature, not only to simple measures of length, area, bulk, weight, and value, but to the measurement of pressure, force, power, specific gravity, temperature, and heat, and in fact all ordinary subjects of quantitative comparison.

One word on the appendices, and I have done.

I have thought it well to append the Decree of the National Convention, August 1793, and the Law of 1795—the foundation-stone and coping-stone respectively of the metric system. Bold in conception and admirable in execution, this great work was marred solely by the unfortunate nomenclature adopted. As evidence of this I have appended a table showing the efforts of the people to get rid of it. Even now, despite its nomenclature, the French system is incomparably superior to our imperial system, while this again is a marked improvement upon the hopeless chaos which prevailed in this country at the beginning of the century. I cannot describe this frightful medley more eloquently than by giving a fairly complete list of weights and measures in actual use and legally recognised before the Act of 1824. Take the stone, the bushel, or the acre, and read the definitions given of them by Parliament itself in 1820.

Lastly, in order to pave the way for further reform on the lines of our own Metric Act, I have appended a rough draught of a Bill for the amendment of the law relating to weights and measures—a Bill which might be brought into Parliament at any moment with or without hope of its passing into law. It would, at any rate, have the effect of familiarising the public mind with a complete metric system, and so pave the way to its eventual adoption. It is quite a mistake to suppose (as our opponents always do) that the English are an exceptionally stupid people, and that they would require a long period of schooling before being able to understand any other measures than those in present use. Italy and Germany transplanted the French system (names and all) on to foreign soil, and the people got used to it in less time than it takes one of our Royal Commissions to publish a report, to the effect that any tampering with existing arrangements would be 'fraught with difficulties and danger.' All such timorous forebodings are unreal and baseless. The whole English people (man, woman, and child) could learn the system advocated in this work in *ten minutes*, and could use it with perfect facility in a week. There is not the slightest necessity to allow the old and new systems to run concurrently for any longer than the time required to substitute new measures for old. As to the cost of this, it is much smaller than is usually supposed. In the case of all metal weights the cost is trifling, and, having regard to the great diminution in the number required, it may be doubted whether there would not be a gain. As to the wooden measures, they are not costly, and as it is they have to be frequently renewed. And as to the difficulty of constructing the new measures, it is perfectly well known that this has already been accomplished. The Board of Trade is now in possession of a fairly complete set, such as I have enumerated in Schedule II. of the Bill. For the convenience of those who wish to use the basic system, I have in Schedule III. given our imperial measures in terms of the basic, and the basic measures in terms of the imperial.

APPENDICES

I.

TABLE OF METRIC WEIGHTS AND MEASURES

(Taken from the Law of April 1795, and Annexed to the Law of July 4, 1887, at present operative)

Measures of Length

Myriamètre	=	10,000 mètres.
Kilomètre	=	1,000 ,,
Hectomètre	=	100 ,,
Mètre	=	Fundamental unit of weights and measures (10,000,000th part of the meridian quadrant).
Décimètre	=	10th of the mètre.
Centimètre	=	100th ,,
Millimètre	=	1,000th ,,

Measures of Land

Hectare	=	100 ares, or 10,000 square mètres.
Are	=	A square 10 mètres each side, or 100 square mètres.
Centiare	=	100th of the are, or 1 square mètre.

Measures of Capacity for Liquids and Dry Goods

Kilolitre	=	1,000 litres.
Hectolitre	=	100 ,,
Décalitre	=	10 ,,
Litre	=	A cubic décimètre.
Décilitre	=	10th of the litre.

Measures of Solidity

Décastère	=	10 stères.
Stère	=	cubic mètre.
Décistère	=	10th of the stère.

Weights

		1,000 kilogrammes, the weight of a cubic mètre of water, and the ton used for ships.
		100 kilogrammes, the metric quintal.
Kilogramme	=	1,000 grammes (weight in a vacuum of a cubic décimètre of distilled water, at the temperature of 4° Centigrade).
Hectogramme	=	100 grammes.
Décagramme	=	10 „
Gramme	=	Weight of a cubic centimètre of water at 4° Centigrade.
Décigramme	=	10th of the gramme.
Centigramme	=	100th „
Milligramme	=	1,000th „

Coins

Franc	=	5 grammes of silver of the standard of $\tfrac{9}{10}$ fine silver.
Décime	=	10th of the franc.
Centime	=	100th „

In conformity with the provisions of the Law of April 7, 1795, each weight and measure of capacity in this decimal scale is to have also its double and its half.

DÉCRET DU 1ᵉʳ AOÛT 1793

La Convention Nationale, convaincue que l'uniformité des poids et mesures est un des plus grands bienfaits qu'elle puisse offrir aux citoyens français :

Après avoir entendu le rapport de son comité d'instruction publique sur les opérations qui ont été faites par l'Académie des Sciences, d'après le décret du 8 mai 1790,

Déclare qu'elle est satisfaite du travail qui a déjà été exécuté par l'Académie

APPENDIX

des Sciences sur le système des poids et mesures ; qu'elle en adopte les résultats pour établir ce système dans toute la France, sous la nomenclature du tableau annexé à la présente loi, et pour *l'offrir à toutes les nations :*
En conséquence, la Convention décrète ce qui suit:—

Art. 1.—Le nouveau système des poids et mesures, fondé sur la mesure du méridien de la terre et la division décimale, servira uniformément dans toute la France.

Art. 2.—Néanmoins, pour laisser à tous les citoyens le temps de prendre connaissance de ces nouvelles mesures, les dispositions de l'article précédent ne seront obligatoires qu'au 1er juillet 1794, les citoyens sont *seulement invités* d'en faire usage avant cette époque.

Art. 3.—Il sera fait, par des artistes au choix de l'Académie des Sciences, des étalons des nouveaux poids et mesures, qui seront envoyés à toutes les administrations de département et de district.

Art. 4.—L'Académie des Sciences nommera quatre commissaires pris dans son sein, et le comité d'instruction publique en nommera deux pour surveiller la construction des étalons ; ils en constateront l'exactitude et signeront les instructions destinées à accompagner les envois qui seront faits par le Ministre de l'Intérieur.

Art. 5.—L'Académie des Sciences enverra au comité d'instruction publique un devis estimatif des frais qu'exigera la construction des étalons, pour que la Convention en puisse décréter les fonds nécessaires.

Art. 6.—Les étalons seront conservés avec le plus grand soin dans un lieu destiné à cet objet, dont la clef restera entre les mains d'un des commissaires de chaque corps administratif.

Art. 7.—Afin d'empêcher la dégradation des étalons, les corps administratifs nommeront, dans chaque chef-lieu de département ou de district, une personne éclairée pour assister à la communication que les artistes prendront de ces étalons, dans la vue de construire des instruments de mesures et de poids à l'usage des citoyens.

Art. 8.—Dès que les nouveaux étalons seront parvenus aux administrations de district, toutes les municipalités de chaque district seront tenues de faire construire des instruments de mesures et de poids qui resteront déposés à la maison commune.

Art. 9.—Le recueil des différents mémoires rédigés jusqu'à présent par les commissaires de l'Académie, qui comprend les détails des opérations faites pour parvenir au nouveau système des poids et mesures, sera imprimé et accompagnera l'envoi des étalons.

Art. 10.—La Convention charge l'Académie de la composition d'un livre à l'usage de tous les citoyens, contenant des instructions simples sur la manière de se servir des nouveaux poids et mesures, et sur la pratique des opérations arithmétiques relatives à la division décimale.

Art. 11.—Des instructions sur les nouvelles mesures et leurs rapports aux anciennes les plus généralement répandues entreront dans les livres élémentaires d'arithmétique qui seront composés pour les écoles nationales.

LOI DU 18 GERMINAL, AN III (1795)

La Convention, voulant assurer au peuple français le bienfait des poids et mesures uniformes et invariables précédemment décrétés, et prendre les moyens les plus efficaces pour en faciliter l'introduction dans toute la France, après avoir entendu le rapport de son comité d'instruction publique, décrète ce qui suit :—

Art. 1.—L'époque prescrite par le décret du 1er août 1793 pour l'usage des nouveaux poids et mesures est prorogée, quant à la disposition obligatoire, jusqu'à ce que la Convention y ait statué de nouveau, en raison des progrès de la fabrication : les citoyens sont cependant invités de donner une preuve de leur attachement à l'unité et à l'indivisibilité de la République, en se servant, dès à présent, des nouvelles mesures dans leurs calculs et transactions commerciales.

Art. 2.—Il n'y aura qu'un seul étalon des poids et mesures pour toute la France; ce sera une règle de platine sur laquelle sera tracé le mètre qui a été adopté pour l'unité fondamentale de tout le système des mesures.

Cet étalon sera exécuté avec la plus grande précision, d'après les expériences et observations des commissaires chargés de sa détermination, et il sera déposé près du Corps Législatif, ainsi que le procès-verbal des opérations qui auront servi à le déterminer, afin qu'on puisse les vérifier dans tous les temps.

Art. 3.—Il sera envoyé dans chaque chef-lieu de district un modèle conforme à l'étalon prototype dont il vient d'être parlé, et, en outre, un modèle de poids exactement déduit du système des nouvelles mesures. Ces modèles serviront à la fabrication de toutes les sortes de mesures employées aux usages des citoyens.

Art. 4.—L'extrême précision qui sera donnée à l'étalon en platine ne pouvant pas influer sur l'exactitude des mesures usuelles, ces mesures continueront d'être fabriquées d'après la longueur du mètre adoptée par les décrets antérieurs.

Art. 5.—Les nouvelles mesures seront distinguées dorénavant par le surnom de *Républicaines*, leur nomenclature est définitivement arrêtée comme il suit :
On appellera :
Mètre, la mesure de longueur égale à la dix millionième partie de l'arc du méridien terrestre compris entre le pôle boréal et l'équateur ;
Are, la mesure de superficie pour les terrains, égale à un carré de dix mètres de côté ;
Stère, la mesure destinée particulièrement au bois de chauffage, et qui sera égale au mètre cube ;
Litre, la mesure de capacité, tant pour les liquides que pour les matières sèches, dont la contenance sera celle du cube de la dixième partie du mètre ;
Gramme, le poids *absolu* d'un volume d'eau pure, égal au cube de la centième partie du mètre, et à la température de la glace fondante ;
Enfin, l'unité des monnaies prendra le nom de *Franc* pour remplacer celui de *Livre* usité jusqu'aujourd'hui.

Art. 6.—La dixième partie du mètre se nommera *décimètre*, et sa centième partie *centimètre* ;
On appellera *décamètre* une mesure égale à dix mètres, ce qui fournit une mesure très-commode pour l'arpentage ;
Hectomètre signifiera une longueur de cent mètres ;
Enfin, *kilomètre* et *myriamètre* seront des longueurs de mille et dix mille mètres, et désigneront principalement les distances itinéraires.

Art. 7.—Les dénominations des mesures des autres genres seront déterminées d'après les mêmes principes que celles de l'article précédent.
Ainsi, *décilitre* sera une mesure de capacité dix fois plus petite que le litre ; *centigramme* sera la centième partie du poids d'un gramme.
On dira de même *décalitre* pour désigner une mesure contenant dix litres ; *hectolitre* pour une mesure égale à cent litres ; un *kilogramme* sera un poids de mille grammes.
On composera d'une manière analogue les noms de toutes les autres mesures.
Cependant lorsqu'on voudra exprimer les dixièmes ou les centièmes d'un franc, unité des monnaies, on se servira des mots *décime* et *centime*, déjà reçus en vertu des décrets antérieurs.

Art. 8.—Dans les poids et les mesures de capacité, chacune des mesures décimales de ces deux genres aura son double et sa moitié, afin de donner à la vente des divers objets toute la commodité que l'on peut désirer ; il y aura donc le *double litre* et le *demi-litre* ; le *double hectogramme* et le *demi-hectogramme*, et ainsi des autres.

Art. 9.—Pour rendre le remplacement des anciennes mesures plus facile et moins dispendieux, il sera exécuté par parties et à différentes époques. Ces époques seront décrétées par la Convention Nationale, aussitôt que les mesures républicaines se trouveront fabriquées en quantités suffisantes, et que tout ce qui tient à l'exécution de ces changements aura été disposé. Le nouveau système sera d'abord introduit dans les assignats et monnaies, ensuite dans les mesures linéaires ou de longueur, et progressivement étendu à toutes les autres.

Art. 10.—Les opérations relatives à la détermination de l'unité des mesures de longueur et de poids, déduites de la grandeur de la terre, commencées par l'Académie des Sciences, et suivies par la commission temporaire des mesures, en conséquence des décrets des 8 mai 1790 et 1er août 1793, seront continuées, jusqu'à leur entier achèvement, par des commissaires particuliers, choisis principalement parmi les savants qui y ont concouru jusqu'à présent, et dont la liste sera arrêtée par le comité d'instruction publique. Au moyen de ces dispositions, l'administration dite commission temporaire des poids et mesures est supprimée.

Art. 11.—Il sera formé, en remplacement, une agence temporaire, composée de trois membres, et qui sera chargée, sous l'autorité de la commission d'instruction publique, de tout ce qui concerne le renouvellement des poids et mesures, sauf les opérations confiées aux commissaires particuliers, dont il est question dans l'article précédent.

Les membres de cette agence seront nommés par la Convention Nationale, sur la proposition de son comité d'instruction publique. Leur traitement sera réglé par ce comité, en se concertant avec celui des finances.

Art. 12.—Les fonctions principales de l'agence temporaire seront—

1°. De rechercher et employer les moyens les plus propres à faciliter la fabrication des nouveaux poids et mesures pour les usages de tous les citoyens ;

2°. De pourvoir à la confection et à l'envoi des modèles qui doivent servir à la vérification des mesures dans chaque district ;

3°. De faire composer ou répandre les instructions convenables pour apprendre à connaître les nouvelles mesures et leurs rapports avec les anciennes ;

4°. De s'occuper des dispositions qui deviendraient nécessaires pour régler l'usage des mesures républicaines et de les soumettre au comité d'instruction publique, qui en fera rapport à la Convention Nationale ;

5°. D'arrêter les états de dépenses de toutes les opérations qu'exigeront la détermination et l'établissement des nouvelles mesures, afin que ces dépenses puissent être acquittées par la commission d'instruction publique ;

6°. Enfin de correspondre avec les autorités constituées et les citoyens dans toute la République sur tout ce qui sera utile pour hâter le renouvellement des poids et mesures.

Art. 13.—La fabrication des mesures républicaines sera faite autant que possible par des machines, afin de réunir à l'exactitude la facilité et la célérité dans les procédés, et par conséquent de rendre l'achat des mesures d'un prix médiocre pour les citoyens.

Art. 14.—L'agence temporaire favorisera la recherche des machines les plus avantageuses ; elle en commandera, s'il en est besoin, aux artistes les plus habiles, ou les proposera au concours, suivant les circonstances. Elle pourra aussi accorder des encouragements en avances, matières ou machines, aux entrepreneurs qui prendraient des engagements convenables pour quelque partie importante de la fabrication des nouveaux poids et mesures, mais, dans tous les cas, l'agence sera tenue de prendre l'autorisation du comité d'instruction publique.

Art. 15.—L'agence temporaire déterminera les formes des différentes sortes de mesures, ainsi que les matières dont elles devront être faites, de manière que leur usage soit le plus avantageux possible.

Art. 16.—Il sera gravé sur chacune de ces mesures leur nom particulier, elles seront, en outre, marquées du poinçon de la République qui en garantit l'exactitude.

Art. 17.—Il y aura, à cet effet, dans chaque district, des vérificateurs chargés de l'apposition du poinçon. La détermination de leur nombre et de leurs fonctions fera partie des règlements que l'agence préparera, pour être ensuite soumise à la Convention Nationale par son comité d'instruction publique.

Art. 18.—Le choix des mesures appropriées à chaque espèce de marchandise aura lieu de manière que, dans les cas ordinaires, on n'ait pas besoin de fractions plus petites que les centièmes.

L'agence recherchera les moyens de remplir cet objet, en s'écartant, le moins possible, des usages du commerce.

Art. 19.—Au lieu des tables de rapports entre les anciennes et les nouvelles mesures, qui avaient été ordonnées par le décret du 8 mai 1790, il sera fait des échelles graphiques pour estimer ces rapports, sans avoir besoin d'aucun calcul.

L'agence est chargée de leur donner la forme la plus avantageuse, d'en indiquer la méthode et de la répandre autant qu'il sera nécessaire.

Art. 20.—Pour faciliter les relations commerciales entre la France et les nations étrangères, il sera composé, sous la direction de l'agence, un ouvrage qui offrira les rapports des mesures françaises avec celles des principales villes de commerce des autres peuples.

Art. 21.—Pour subvenir à toutes les dépenses relatives à l'établissement des nouvelles mesures, ainsi qu'aux avances indispensables pour le succès de cette opération, il y sera affecté provisoirement un fonds de cinq cent mille livres que la trésorerie nationale tiendra, à cet effet, à la disposition de la commission d'instruction publique.

Art. 22.—La disposition de la loi du 24 Novembre 1793, qui rend obligatoire l'usage de la division décimale du jour et de ses parties, est suspendue indéfiniment.

Art. 23.—Les articles des lois antérieures au présent décret, et qui y sont contraires, sont abrogés.

Art. 24.—Aussitôt après la publication du présent décret, toute fabrication des anciennes mesures est interdite en France, ainsi que toute importation des mêmes objets venant de l'étranger, à peine de confiscation et d'une amende du double de la valeur des dits objets.

La commission des administrations civiles police et tribunaux, et celle des revenus nationaux sont chargées de l'exécution du présent article.

Art. 25.—Dès que l'étalon prototype des mesures de la France aura été déposé au Corps Législatif par les commissaires chargés de sa confection, il sera élevé un monument pour le conserver et le garantir de l'injure des temps.

L'agence temporaire s'occupera d'avance du projet de ce monument, destiné à consacrer, de la manière la plus indestructible, la création de la République, les triomphes du peuple français et l'état d'avancement où les lumières sont parvenues dans son sein.

Art. 26.—Le comité d'instruction publique est chargé de prendre tous les moyens de détail nécessaires pour l'exécution du présent décret et l'entier renouvellement des poids et mesures dans toute la France.

Il proposera successivement à la Convention les dispositions législatives qui devront en dépendre.

Art. 27.—L'agence temporaire rendra compte de ses opérations à la commission d'instruction publique et au comité de ce nom, avec lequel elle pourra correspondre directement pour la célérité des opérations.

Art. 28.—Il est enjoint à toutes les autorités constituées, ainsi qu'aux fonctionnaires publics, de concourir, de tout leur pouvoir, à l'opération importante du renouvellement des poids et mesures.

Progressive Nomenclature of French System

Académie des Sciences		Commission Temporaire, 1793	Loi du 18 Germinal, 1795	Arrêté du 18 Brumaire, An IX, 1801
Décade				
Degré	. . . .	Grade ou degré		
Poste	. . . .	. . .	Myriamètre	Lieue
Mille	. Millaire	Millaire	Kilomètre	Mille
Stade	. . . .	. . .	Hectomètre	
Perche	. . . .	. . .	Décamètre	Perche
Mètre	. Mètre	Mètre	Mètre	Mètre
Palme	. Décimètre	Décimètre	Décimètre	Palme
Doigt	. Centimètre	Centimètre	Centimètre	Doigt
Trait	. Millimètre	Millimètre	Millimètre	Trait
Tonneau	. Muid	Cade	Kilolitre	Muid
Setier	. Décimuid	Décicade	Hectolitre	Setier
Boisseau	. Centimuid	Centicade	Décalitre	Boisseau, Velte
Pinte	. Pinte	Cadil	Litre	Pinte
			Décilitre	Verre
			Centilitre	
Millier	. Millier	Bar	. . .	Millier
Quintal	. . . .	Décibar	. . .	Quintal
Décal	. . . .	Centibar	Myriagramme	
Livre	. Grave	Grave	Kilogramme	Livre
Once	. Décigrave	Décigrave	Hectogramme	Once
Drame	. Centigrave	Centigrave	Décagramme	Gros
Maille	. Milligrave	Gravet	Gramme	Denier
Grain	. . . .	Décigravet	Décigramme	Grain
		Centigravet	Centigramme	
		Milligravet	Milligramme	
		Are	Hectare	Arpent
		Deciare		
		Centiare	Are	Perche carrée
			Centiare	Mètre carré
			Stère [1]	Stère
			Décistère	Solive
Unité monétaire [2]	.			
			Franc [3]	Franc
			Décime	Sol
			Centime	Denier

[1] Mètre cube.   [2] Décagr. d'argent.   [3] 5 grammes d'argent.

II.

ENGLISH MEASURES IN 1800

(GATHERED FROM OLD PARLIAMENTARY REPORTS)

ACRE, ENGLISH: A measure of land containing 4 roods = 160 perches = 4,840 square yards = 43,560 square feet.
SCOTCH: 6,150 square yards.
IRISH: 7,840 square yards.
CUSTOMARY ACRES:
Bedfordshire: Sometimes 2 roods.
Cheshire: Formerly and still in some places, 10,240 square yards.
Cornwall: Sometimes one of the Welsh acres of 5,760 yards.
Dorsetshire: Generally 134 perches.
Hampshire: From 107 to 120 perches, but sometimes 180.
Herefordshire: Two-thirds of a statute acre of hops, about half an acre, containing 1,000 plants.
of wood, 256 perches.
Leicestershire: 2,308¾ square yards.
Lincolnshire: 5 roods, particularly for copyhold land.
Staffordshire: Nearly 2¼ acres.
Sussex: 107, 110, 120, 130, or 212 perches.
Short acre, 100 or 120 perches.
Forest acre, 180 perches.
Westmorland: 6,760 square yards, or 160 perches of 6½ yards square; in some parts the Irish acre is used.
Worcestershire: Hop acre, 1,000 stocks, or 90 perches; sometimes 132 or 141 perches.
N. Wales: Erw, or true acre, 4,320 square yards; stang or customary acre, 3,240 square yards.

ACRE: as a measure of length.
Bedfordshire: } a chain of 4 poles,
Buckinghamshire: } or 22 yards.
Derbyshire: 4 'roods,' each of 7 or of 8 yards.
Yorkshire: 28 yards.
ANKER: Ten wine gallons.
Scotland: Twenty Scotch pints.
AUME or AWM: A tierce of wine, or 42 gallons.

BARLEYCORN: ⅓ inch.
BARREL:
of ale and beer, 36 gallons, 43 G. 3. Before this Act the legal barrel of ale was only 32 gallons in London, of beer 36; and in the country both were 34. 23 H. 8.
of anchovies, 16 pounds. 27 G. 3.
of apples, 3 bushels. 12 Car. 2.
of barilla, 2 cwt. 12 Car. 2.
of beef, 32 wine gallons. 38 G. 3.
of candles, 10 dozen pounds. 12 Car. 2.
of butter, 224 pounds; but 106 only of Essex butter; 150 of Suffolk.
of coals, 3 Winchester bushels. 18 G. 3.
of cod fish, wet, 32 gallons. 5 G. 1.
of eels, 42 gallons. 32 Ed. 4; but 30 by 2 H. 6.
of fish in general, to be gauged by wine measure. 38 G. 3.
of gunpowder, 100 pounds neat. 4 G. 2.
of herrings, 42 gallons. 32 Ed. 4. and 5 G. 1; but 30 gallons 2 H. 6. Estimated to contain 235 pounds of

APPENDIX 205

BARREL:
fish if packed with small salt; 212 if with great salt; and 4 unpacked are considered as equal to 3 packed. 26 G. 3.
of honey, 32 wine gallons. 23 Eliz. Otherwise 42 gallons of 12 pounds each.
of mum, 42 gallons. 11 G. 1. Otherwise 32.
of nuts, 3 bushels. 12 Car. 2.
of oil, 31½ wine gallons.
of pilchards, or mackerel, salted, 50 gallons. 41 G. 3.
of pork, to be gauged by wine measure. 38 G. 3.
of potashes, 2 cwt. 12 Car. 2.
of raisins, 1 cwt.
of salmon, 42 gallons. 32 Ed. 4; 5 G. 1.
of soap, 256 pounds. 10 Anne.
of spirits, 31½ wine gallons.
of spruce beer, 42 gallons. 11 G. 1.
of tar, 31½ gallons. 24 G. 2.
of vinegar, 34 ale gallons. 10 and 11 W. 3.
of wine, 31½ gallons. 1 Rich. 3.
Isle of Man: of lime, 6 Winchester bushels.
Guernsey and Jersey: of charcoal and lime, 120 pots = 60 gallons.
Wales:
of lime, in some counties 3 provincial bushels of 10 gallons each, making 3¾ Winchester bushels, the measure being square and without a bottom.
of culm, in some parts 4 heaped bushels; in some 40 gallons.
Scotland:
of beef or herrings, 32 gallons.
of salmon, 42 gallons.
Argyleshire: of herrings, 32 gallons English.
Banffshire: of coal, about 12 stone. 210 pounds English.
Cromarty and Ross-shire:
of coal, 10 gallons Scotch.
of limestone, 32 gallons English.
Galloway: of coal, 85 gallons English.
Kincardineshire: of flax, 18 pecks.
Moray and Nairn: of coal, 3 Winchester bushels heaped, weighing about 13 stone Dutch.

BARREL:
Ireland:
of grain, in general, formerly 4 bushels of 10 gallons each.
of barley, 16 stone, 224 pounds.
of beans, pease, and wheat, 20 stone.
of malt, 12 stone.
of oats, 14 stone.
of potatoes, 20 stone.
of roche lime, 40 gallons of $217\frac{7}{6}$ cubic inches each.
BASKET:
of medlars, 2 bushels. 12 Car. 2.
of Spa water, 150 flasks. 11 G. 1.
Kent: of cherries, 48 pounds.
BAT: S. Wales: a perch of 11 feet square.
BAY: Derbyshire: of slater's work, 500 square feet.
BILLET: of firewood, 3 feet 4 inches long; if single, 7¼ inches about; cast, 10, two cast, 14. 43 Eliz. Altered to 11, 12½, 13, and 15 inches about. 9 Anne.
BIND:
of eels, 10 sticks = 250 eels. 31 Ed. 1.
of skins, 32; of some kinds, 40. 31 Ed. 1.
BING: Durham and Northumberland: of lead ore, 8 cwt.
BOLL: of canvas, 28 ells = 35 yards. 12 Car. 2.
Durham: 2 bushels.
Alnwick:
of barley and oats, 6 bushels.
of wheat, 2 bushels.
Hexham:
of barley and oats, 5 bushels.
of pease, rye, and wheat, 4 bushels.
Newcastle: 2 bushels.
Wooler: 6 bushels.
Westmorland: of rye, 2 bushels.
Isle of Man:
of barley and oats, 6 bushels.
of peas, 4 bushels.
of potatoes, 16 heaped pecks.
of wheat, 4 bushels, weighing 64 pounds each.
Scotland:
of grain, the boll is the legal measure, at least recognised, 24 G. 2; 18 G. 3. It contains 4 firlots, nearly 6 Winchester bushels, or more accurately, 5·9626.
of oatmeal, 140 pounds English, 128 Scotch

BOLL:
Scotland:
Troy. 43 G. 3. That is, 8 Dutch stone, or 10 English, 16 pecks. First established by law as a boll in 1696.
of marl, 8 cubic feet.
But nearly every county in Scotland had its own customary boll, which also varied according to the commodity sold.
BOLT or BOULT: of oziers.
Berkshire: a bundle, measuring 42 inches round, 14 inches from the butts.
Essex: a bundle, of which 80 make a load.
Hampshire: 42 inches round at the lower band.
BOTTLE:
of aqua fortis, 4 gallons. 8 G. 1.
of wine, about 5 to a gallon. 43 G. 3.
BOX: of aloes, 14 pounds.
Derbyshire: of coals, at the pit 2½ bushels, striked.
Durham: of salmon, 8 stone.
BREAD WEIGHT: Essex: Troy weight.
BUCKET: Buckinghamshire and Hertfordshire: of chalk, 1½ bushel.
BUNCH:
Cambridgeshire:
of oziers, a bundle 45 inches round at the band.
of reeds, a bundle 28 inches round, formerly an ell.
Essex: of teazles, 25 heads; otherwise a glean.
Gloucestershire: of teazles, 20; a glen; of king's teazles, 10.
Yorkshire, N.R.: of teazles, 10
BUNDLE:
of bast ropes, 10. 12 Car. 2.
of brown paper, 40 quires. 11 G. 1.
Berkshire:
of hoops, 120 to 480, according to the sort.
of birch brooms, 1 or 2 dozen, according to the size.
Devonshire:
of barley straw, 35 pounds.
of oat straw, 40 pounds.
of wheat straw, 28 pounds.
Gloucestershire:
of hogshead hoops, 36.
of oziers, about 1¼ round, or a little more.

BUNDLE:
Hampshire:
of oziers, 42 inches round the lower band.
of rafter poles, 100 hoops.
Surrey: of hoops, 60.
BURDEN: of steel, 180 pounds.
BUSHEL:
of coals, to contain a Winchester bushel and a quart of water; to be 19½ inches in diameter from outside to outside, and to be heaped in the form of a cone 6 inches high. 12 Anne.
of fruit, or water measure, was to consist of 4 heaped pecks, 33 dry quarts.
Equivalent weight:
of meal or wheat, ½ cwt. 14 G. 3.
of wheat, a bushel is estimated at 57 pounds. 29 G. 3.
of rye, 55; of barley, 49; of oats, 38. 31 G. 3.
of biscuit, bread, or flour, 42 pounds. 14 G. 3. Of flour, 45 pounds. 31 G. 3.
of salt, 56 pounds. 5 W. & M.; 38 G. 2. Of rock salt, 128 pounds. 5 W. & M.; then 75, 10 and 11 W. 3; and 65, 1 Anne; 38 G. 2. Of foreign salt, 84 pounds. 1 Anne; perhaps superseded by 38 G. 2.
of wheat meal, 56 pounds. 31 G. 3.
In IRELAND: a bushel is to weigh, of wheat, 70 pounds; of barley, 42; of oats, 88½; of malt, 35. 1. A. 7 G. 2; 11 G. 2.
Equivalent quantity:
of barley, makes 49 pounds of barley-meal, 31 G. 3; 48 pounds, 41 G. 3.
of oats, makes 22 pounds of oat-meal, 31 G. 3; 42 pounds, 41 G. 3.
of rye, makes 55 pounds of rye-meal. 31 G. 3.
of wheat, makes 70 pounds of bread or biscuit. 31 G. 2.
Bedfordshire: 2 pints above the standard.
Berkshire: of corn, in some parts 9 gallons.
Cheshire:
of barley, 60 pounds.
of oats, 45 to 50 pounds.
of potatoes, 90 pounds.
of wheat, 70 to 75 pounds.
Cornwall: 24 gallons. The double measure

BUSHEL:
Cornwall:
  of 16 gallons is also used in the eastern parts, and runs occasionally to 17 or 17½: the triple in the western.
  of potatoes, 220 pounds.
Cumberland. Carlisle: 96 quarts = 24 gallons.
Penrith:
  of barley, oats, and potatoes, 20 gallons.
  of rye and wheat, 16 gallons.
Derbyshire: of potatoes, often 90 pounds.
Devonshire:
  of barley, often 50 pounds.
  of oats, often 30 or 40 pounds.
  of wheat, the fourth peck heaped.
Dorsetshire: of hemp seed, sometimes 9 gallons.
Durham: of corn, generally 5 per cent. above the standard; in some parts 8½ gallons.
Stockton: of oats, 35 pounds.
  of wheat, 60 pounds.
Gloucestershire: commonly 9½ gallons, but varying from 9 to 9¼ and 10.
Herefordshire:
  of grain, 10 gallons.
  of malt, 8½ gallons.
Lancashire: of potatoes, generally 90 pounds not cleaned.
Liverpool: of barley, beans, and oats, 34½ customary quarts, making 9 gallons Winchester measure; 21 bushels sold for 20.
Leicestershire:
  of grain, 8½ to 9 gallons.
  of malt, 8 gallons.
  of potatoes, 80 pounds.
Middlesex: of potatoes, 56 pounds.
Northumberland: the Winchester bushel, variously subdivided.
Oxfordshire: of wheat, 9 gallons, 3 pints.
Shropshire:
  of barley, pease, and wheat, 9½ to 10 gallons.
  of wheat, weighing from 70 to 80 pounds.
  of oats, at Shrewsbury, 3½ bushels, weighing about 93 pounds. In other parts variously heaped.
Somersetshire: of coal, at the pits, 9 gallons.

BUSHEL:
Staffordshire. Wolverhampton:
  of barley, beans, oats, and pease, 9½ gallons.
  of wheat, 72 pounds.
Surrey:
  of potatoes, 60 pounds.
  of turnips, 50 pounds.
Sussex: of wheat, in some parts 9 gallons.
Westmorland, 3 Winchester bushels.
Appleby: of barley, 2½ bushels.
  of potatoes, 2 bushels.
Worcestershire: at Worcester, 8½ gallons; at Evesham, 9; in some parts 9½ or 9¾. Of wheat, 9 gallons weigh 70 pounds, and make 56 of flour.
Yorkshire:
  E.R.: farmers sell by bushel above the standard; corn merchants by the Winchester bushel.
  N.R.: in the southern part 1 quart above the standard, in the northern 2; sometimes 10 per cent., or more than 3.
North Wales. Anglesey: of potatoes, 74 pounds.
South Wales:
  of oats, the Winchester bushel of the old kind of oats required to weigh 41½ pounds; of the new 45 pounds.
  of coal, ¾ of a cwt. = 84 pounds.
Brecknockshire, 10 gallons. See Radnorshire.
Monmouthshire, from 10 to 10½, and nearly 11 gallons.
Montgomeryshire, 20 gallons, called 2 strikes.
Welshpool:
  of malt, $\frac{1}{10}$ of the corn bushel = 18 gallons.
  of oats, 7 hoops of 5 gallons, heaped.
Radnorshire and *Brecknockshire*.
Fishguard, 2 Winchester bushels.
Caerphilly: of wheat, the Winchester bushel, estimated to weigh 67¾ pounds; at Aberthaw, 64; at other places the bushel of 10 gallons is required to weigh 80 pounds.
Guernsey, 6 gallons:
  of wheat, to weigh 38 pounds English.
  of barley, 50 pounds = 54½ English.
Scotland: In Aberdeen there is a brass

BUSHEL:
  Scotland:
    standard bushel of Q. Anne, 1707, which contains 13 cubic inches less than the Winchester standard. A bushel used in the county contains 40 cubic inches less.
  Ayrshire, 2 pecks.
  Galloway:
    of barley, from 46 to 58 pounds.
    of lime, the Carlisle bushel = 3 Winchester bushels.
BUTT:
  of wine or cider, 2 hogsheads. 5 Anne.
  of beer, in London, 3 hogsheads.
  of salmon, 84 gallons. 2 H. 6.

CABOT: *Jersey*: of wheat, $\frac{1}{12}$ of an English bushel, weighing about 34½ pounds English, the small cabot; 4 of which make 3 large cabots, used for barley, and all corn except wheat.
CADE:
  of herrings, 500. 12 Car. 2.
  of sprats, 1,000.
CARAT: of the jewellers, 4 grains.
CARRIAGE: of lime, 64 heaped bushels.
CART: of coals, 8¾ cwt. 6 & 7 W. 3. At Liverpool, 1 chaldron.
CART LOAD, from 3 tons to 27 cwt. according to the wheels and the season. 13 G. 3.
CASK:
  of almonds, 3 cwt.
  of cloves, mace, or nutmegs, about 300 pounds neat. 6 G. 1.
  of pilchards, 50 gallons. 5 G. 1.
  of tobacco, cask or chest of 224 pounds. 10 & 11 W. 3. At least 450 pounds. 24 G. 2.
  of wheat flour, 2 cwt.
  Gloucestershire: of cider, usually 110 gallons.
  Scotland: of meal or of salt fish, measured by the wine gallon.
  Caithness: of butter, from 72 to 84 pounds.
CAST: of earthen or stone pots, 8 gallons. 12 Car. 2.
CHAIN: 4 poles = 22 yards = 66 feet.
  Scotland, 74·4 feet; sometimes improperly 74 only, ten square chains making an acre, as in England.

CHALDER:
  Scotland, for chaldron, called 4 quarters or 4½, but equal to nearly 12 quarters Winchester measure.
  of corn, 16 bolls.
  Dumbartonshire:
    of lime, 64 bushels.
    of lime shells, 32 bushels.
  Renfrewshire:
    of lime, 32 bushels.
    of lime shells, 16 bushels.
  Stirlingshire: of lime, in some places 24 firlots, each of 29 Scotch pints.
CHALDRON: of coal, 4 vats, 36 coal bushels. 16 & 17 Car. 2. At Newcastle, 8 wains, or 52½ cwt. estimated at 53 in the contents of boats. 6 & 7 W. 3. In London, about half.
  Cambridgeshire: of lime, 40 bushels.
  Derbyshire: of lime, in some parts, 82 heaped bushels.
  Lincolnshire: of coals, 48 bushels.
  Surrey: of lime, 32 bushels.
  Yorkshire, E.R.:
    of coals, at Bridlington, 48 bushels.
    of lime, 32 bushels.
  N.R.: of coals and lime, 32 bushels.
  North Wales: of coals, 32 bushels.
  Guernsey and *Jersey*: of coals, 72 bushels.
CHEEF: of fustian, 13 yards. 31 Ed. 1.
CHOPIN or CHOPPIN: *Scotland*, ½ a pint, 2 mutchkins = 52½ cubic inches, about 2 English pints.
CLOVE: 7 pounds: of cheese, 7 pounds. 9 H. 6. But sometimes 8.
CLUE: of yarn or hemp, 4,800 yards, making 2 or 3 skains.
COOM or COOMB: half a quarter = 4 bushels.
CORD: a measure for wood, properly a double cube of 4 feet = 128 cubic feet.
  Derbyshire: 128, 155, or 162½ cubic feet.
  Sussex: 14 × 3 × 3 feet = 126 cubic feet.
CORF or CORVE: of coal: *Durham*, 4 bushels, or 3¼ cwt.
  Derbyshire: 2¼ level bushels, or 2 cwt.; called also basket, box, or tub.
CRAN: of herrings, 34 wine gallons. 41 G. 3.

DACRE or DICKER: a decad. 31 Ed. 1.
  of razors. 12 Car. 2.
  of gloves, 10 pair.

APPENDIX

DACRE or DICKER:
of horse-shoes, 20 shoes.
DAUGH: *Scotland*: an ancient measure of land, of 48 bolls.
DISH: *Derbyshire*: of lead ore, 14 pints of 48 cubic inches, making 672, in the Low Peak hundred, weighing 58 pounds; but in the High Peak 16 pints.
DOZEN:
of bristles, 12 pounds. 38 G. 3.
of calves' skins, 36 pounds. 12 C. 2.
of iron, 6 pieces. 31 Ed. 1.
Derbyshire:
of charcoal, 72 local bushels.
of ironstone, 50 cubic feet stacked.
Durham: of poles for lead mines, from 10 to 100, according to the size.
DRACHM or DRAM: in pharmacy, $\frac{1}{8}$ of an ounce Troy = 3 scruples = 60 grains. In general commerce, $\frac{1}{16}$ of an avoirdupois ounce.
DRACHM: by measure; more properly FLUIDRACHM, $\frac{1}{8}$ of a wine pint, $\frac{1}{8}$ of a fluid ounce.
DROP: sometimes a weight of 30 grains, or $\frac{1}{2}$ a drachm, apothecaries' weight; sometimes 27$\frac{1}{2}$, or 1 drachm avoirdupois.
DROP: as a measure:
of water, about $\frac{1}{60}$ of a fluid drachm.
of tinctures, about $\frac{1}{120}$ of a fluid drachm.

ELL: 5 quarters = 45 inches, in England, and legally in Scotland. 43 G. 3.
Shropshire: of linen cloth, 6 quarters = 54 inches.
Jersey: 4 feet = 48 inches.
SCOTLAND, the standard is 37 inches = 37$\frac{1}{2}$ English.
Aberdeenshire: of plaiding, 38$\frac{1}{12}$ inches.
Cromartyshire: 38 inches.
Dumfries-shire: in some parts, 39 inches.
Edinburgh: 37 inches.
of plaiding and stuffs, 39$\frac{1}{2}$ inches.
of land, 37$\frac{1}{2}$ inches.
Nairnshire and *Ross-shire*: 38 inches.
North Wales: of cloths and cottons, the llathen gyvolin, 9 feet.
Guernsey and *Jersey*: 4 feet.
ERW: *South Wales*: a measure of land varying from a little more than 1 to more than 2 acres, containing 4 stangell or cyvar, each generally of 160 perches.

FAGGOT:
of wood, 3 feet long, 24 inches round. 43 Eliz.
of steel, 120 pounds.
FALL: *Aberdeenshire* and elsewhere: of land, 6 ells square, $\frac{1}{160}$ of an acre, as the perch is of the English acre.
FAN: *Cambridgeshire*: of chaff, 3 heaped bushels.
FAT or VAT: 9 bushels.
of bristles, 5 cwt.
FATHOM: 2 yards = 6 feet.
FIRKIN: $\frac{1}{4}$ of a barrel. See Barrel.
of butter, 56 pounds. 36 G. 3.
of soap, 64 pounds.
of potatoes; at *Shields*, 2$\frac{1}{4}$ bushels; at *Sunderland*, 3.
FIRLOT: *Isle of Man*: $\frac{1}{2}$ a boll.
SCOTLAND: 31 Scotch pints, or nearly 1$\frac{1}{2}$ Winchester bushels: other measures give 3,208 cubic inches, used for barley, bear, malt, and oats.
of wheat, 21$\frac{1}{4}$ pints = 2,197$\frac{1}{2}$ cubic inches, about 2 per cent. more than a Winchester bushel; other measures make it 2,129 cubic inches, used for beans, pease, rye, white salt, and wheat.
Most of the Scotch counties have also their customary firlots.
FLASK: of Pyrmont water, 3 pints, wine measure. 11 G. 1.
FODDER or FOTHER: of lead, a ton = 20 cwt. 12 C. 2. With miners, 22$\frac{1}{2}$ cwt.; with plumbers, 19$\frac{1}{2}$.
Derbyshire: mill fodder, at the smelting houses, 2,820 pounds; when shipped at Stockwith-on-Trent, 2,408.
Hull: 2,340 pounds.
London: 2,184 = 19$\frac{1}{2}$ cwt.
Northumberland:
of dung and lime, a two-horse cart load.
of pig lead, 21 cwt.; at *Newcastle*, sometimes 22 cwt.
FOOT: 12 inches, $\frac{1}{3}$ of a yard. See Inch.
Durham: of grindstone, 8 inches.
Wales: the ancient Welsh foot is said to have been 3 hand-breadths = 9 inches.

P

Foot:
Scotland: 1·005405 foot English = 12·064864 inches; $\frac{1}{185}$, or $\frac{1}{2}$ per cent. above the English standard.
of the glaziers, commonly 8 inches.
Forpet or Forpit: *Northumberland, Alnwick*: the fourth part of a peck, about 3 quarts.
Hexham: 4 quarts, $\frac{1}{4}$ peck of wheat, $\frac{1}{8}$ of barley and oats.
Wooler: 4 quarts, $\frac{1}{2}$ peck, $\frac{1}{8}$ bushel.
Scotland: the fourth part of a peck, otherwise called a lippie.
Fotmal: of lead, 70 pounds. 31 Edw. 1.
Furlong: $\frac{1}{8}$ of a mile = 10 chains = 220 yards. Square, 10 acres.

Gallon: old standard, 28 making 32 wine measure. 13 Eliz.
Wine gallon: 231 cubic inches, or a cylinder 7 inches in diameter, and 6 high. 5 Anne. Used for beef, pork, and fish. 38 G. 3; and at *Leicester*, for all liquids.
of brandy, estimated to weigh 7 pounds 13 ounces. 32 G. 2.
Winchester gallon: estimated at 272$\frac{1}{4}$ cubic inches. 45 G. 3. *See* Bushel.
of honey, 12 pounds, at the Custom House.
of train oil, 7$\frac{1}{2}$ pounds, at the Custom House.
Guernsey: 252 cubic inches.
Scotland: 8 pints = 840 cubic inches.
Ireland: 217$\frac{6}{10}$ cubic inches, for all purposes. 9 G. 2; 26 G. 3. Irish Acts.
Garb: of blocks or steel, 30 pieces. 31 Ed. 1.
Gaun or Gawn: *Shropshire* and *Wales*; a corruption of gallon, applied to butter, containing 12 pounds.
Bridgnorth: of butter, 16 pounds.
Gill: $\frac{1}{4}$ of a pint, whether of wine or of ale measure.
Scotland: $\frac{1}{16}$ of a pint, 6$\frac{3}{4}$ cubic inches; about an English gill.
Banffshire: $\frac{1}{2}$ a pint English.
Glene: of herrings, 25. 31 Ed. 1.
Goad: of cloth: seems to be 4 feet. 28 Ja. 1. '15*d*. the yard, or 20 the goad.'
of land: *Dorsetshire*, 15 feet 1 inch; called also lug.
Grain: $\frac{1}{7000}$ part of a pound avoirdupois.

Grocé or Gross: commonly 12 dozen.
of bracelets or necklaces, 12 dickers or bundles of 10, make a small gross. 12 C. 2.
of pill boxes, 12 dozen nests of 4 boxes each. 11 G. 1.
Gwaith Gwr: *North Wales*: of peat, 150 square feet, or 50 yards in length, 3 bricks or spits in breadth.

Hand: 4 inches. 32 H. 8; 36 G. 3.
Hank:
of cotton; *Derbyshire*, 840 yards.
of woollen or worsted yarn, 7 raps or leas, each 50 threads of a yard, or a two yard reel in the northern counties. 17 G. 3; or other reels in other places. 24 G. 3; 25 G. 3; 31 G. 3.
Haybands: not to exceed 5 pounds in weight. 31 G. 3.
Head or Chief:
of linen, 10 yards. 31. E. 1.
of hemp, 4 pounds.
Heap: of limestone, in some parts of *Scotland*, 4$\frac{3}{4}$ cubic yards, weighing 4 tons.
Hide: of land, 100 acres.
Hobaid or Hobed: of lime: *South Wales*, 4 pedwran, or quarters, of 5 or 6 quarts each; equal to a peccaid; sometimes to $\frac{1}{2}$ a peccaid of corn.
Anglesea and *Caernarvonshire*: 2 storeds = 4 bushels.
Denbigh, *Flint*, and *Merionethshire*: 2$\frac{1}{2}$ bushels W. In the Vale of *Clwyd* and part of *Flintshire*: 21 hobeds are sold for a score.
Montgomeryshire: from 2$\frac{1}{2}$ to 2$\frac{3}{4}$ bushels.
Hogshead:
of ale or beer, 1$\frac{1}{2}$ barrel. Formerly of ale, 48 gallons; of beer, 54.
of molasses, 100 gallons. 22 G. 2.
of wine, 63 gallons. 2 H. 6.
Cornwall: of oats, 9 Winchester bushels.
Devonshire: of lime, sometimes 36 level pecks, or 40; sometimes 11$\frac{1}{2}$ heaped bushels Winch.
Dorsetshire: of lime, 4 bushels.
Herefordshire and *Worcestershire*: of cider, 110 gallons.
Guernsey and *Jersey*: of cider, 120 pots, 60 gallons.

HUNDRED:
of balks, deals, eggs, oars, spars, and stone, 120.   12 C. 2.
of mullets, 8 score = 160.   31 Ed. 1.
Bedfordshire: of faggots, 6 score.
Buckinghamshire: of faggots, 6 score.
Cambridgeshire: of bunches of reeds, 6 score.
Essex:
of faggots and hop-poles, 6 score.
of lime, 25 bushels.
Hampshire: of bavins and faggots, 120, the long hundred.
Northamptonshire: of faggots, 120.
Isle of Man: of herrings, 124.
Dumbartonshire: of herrings, 6 score.
Fifeshire: of herrings, 132.
Roxburghshire and Selkirkshire: of sheep or lambs, sometimes 106.
By measure:
of hempen and linen cloth, 120 yards. 31 Ed. 1.
of lime; Essex, 25 bushels. In some other places 35 bushels, heaped.
WEIGHT or HUNDRED WEIGHT: (cwt.), 112 pounds = 4 quarters = 8 stone; but of aloes, angelica, annotto, asafœtida, bugle, camboge, capers, cotton, down, gentian, ginseng, gum copal, gum guaiacum, indigo, isinglass, manna, myrrh, long pepper, pimento, plums, saccharum saturni, sarsaparilla, tobacco, turmeric, verdigris, and raw linen yarn, 100 pounds are to be reckoned a hundred weight. 38 G. 3;
of yarn, and of some other articles, 6 score pounds. 12 C. 2; 27 G. 3.
of claphoult or clapboards, the small hundred 120 pounds, the great hundred 24 small hundred. 12 C. 2.
Cambridgeshire: of cheese, 120 pounds.
Cheshire: of cheese and hay, 120 pounds, the long hundred.
Derbyshire:
of cheese, 120 pounds.
of coals, on canals, 120 pounds.
of dung, hay, and straw, in some places, 120 pounds.
of lead ore, 120 pounds.
Essex: of potatoes, 120 pounds.
Huntingdonshire: of Leicester cheese, 120 pounds.

WEIGHT or HUNDRED WEIGHT:
Kent: of filberts, 104 pounds.
Lancashire: 100, 112, or 120 pounds.
Leicestershire: of cheese, 120 pounds.
Bridgnorth:
of cheese, 113 pounds; Shrewsbury, 121 pounds.
of coals, at some pits, 120 pounds.
Wolverhampton: of cheese, 120 pounds.
Jersey: 100 pounds of Rouen, about 110 English.
HUTCH: Renfrewshire: of pyrites, or copperas stone, 2 cwt.
HYLE: Hampshire: of flax, 10 sheaves.

INCAST: Roxburgh and Selkirk, a pound in a stone of wool, and a fleece in a pack, usually given above measure.
INGRAIN: ¼ of a chaldron of coals given above 5 chaldron, in London.
INCH: the length of a pendulum vibrating seconds of mean solar time in a vacuum on the level of the sea, in the latitude of London, is 39·1393 inches.
Scotland : = 1·0084084 English inch.

JAR:
of Lucca oil, 25 gallons.
of green vinegar, 100 pounds.
of wheat, 52 pounds.

KEEL: of coals; Newcastle, 8 Newcastle chaldrons = 21 tons 4 cwt. = 424 cwt.
KEMPLE: of straw: Mid-Lothian, windlens, from 14 to 16 stone trone.
KENNING: Durham and Northumberland, ½ a bushel = 2 pecks.
KIBIN: Anglesey and Carnarvon, ¼ a bushel = 2 pecks.
KILDERKIN: half a barrel. See Barrel.
of ale and soap, 18 gallons. 23 H. 8.
of butter, 1 cwt. neat. 13 & 14 C. 2.
KINTAL or QUINTAL: 100 pounds, sometimes 120 pounds.
KISHON: Isle of Man, a peck.
KIVER: Derbyshire, of corn, 12 sheaves.

LAST: 12 sacks. 31 Ed. 1.
of ashes, codfish, pitch, tar, and wheat, 12 barrels. 12 C. 2.
of herrings, 10,000 each containing

LAST:
  1,000 of 6 score each. 31 Ed. 1;
    31 Ed. 3.
  of herrings packed, 12 barrels; unpacked,
    18 barrels.
  of red herrings, 10,000, or 20 cades.
  of pitch and tar, 12 barrels of 31½ gallons
    each. 38 G. 3.
  of skins, '20 dacres' = 120. 31 Ed. 1.
  of stockfish, 1,000. 12 C. 2.
  of butter and soap, 12 ale barrels.
  of corn and seed, 10 quarters.
  of feathers and flax, 1,700 pounds.
  of gunpowder and raisins, 24 barrels.
  of oatmeal and potash, 12 barrels.
  of wool, 24 wey = 42 cwt.
  Cambridgeshire: of oats, 21 comb = 10½
    quarters.
  Huntingdonshire:
    of grain and seeds, 10½ quarters = 84
      bushels.
    of oats, 1½ ton.
  Lincolnshire: Boston, 10½ quarters.
  Yorkshire, N.R.: of rape seed, 10 quarters.
  Scotland: 384 pounds.
  In a ship's burden: 12 tons.
LEAGUE: 3 miles.
  Nautical or Geographical League: $\frac{1}{20}$ of a
    degree of latitude, that is, about 6,076
    yards.
LINE: a term not definite in its sense, being
  sometimes employed for $\frac{1}{12}$, and some-
  times for $\frac{1}{10}$, of an inch.
LINK: $\frac{1}{100}$ of a chain of 66 feet, 7$\frac{92}{100}$ inches.
LIPPIE: *Scotland*: a quarter of a peck = ·0332
  Winchester bushel, nearly a quarter
  and a half of an English peck.
LISPOUND: *Shetland*: 32 pounds English;
  formerly 24 pounds Dutch = 26¼ English.
LLATH: *South Wales*: of land, sometimes
  21 feet square, 160 making an erw;
  sometimes 11½ feet square, 768 to the
  erw; sometimes 24 feet square, 160 to
  the erw. In Anglesey, 5½ llathen make
  an acre of 3,240 square yards, each containing 80 perches of 13½ feet square.
LLATHEN GYVELIN: *Wales*: of cloths, the
  old ell of 9 feet English.
LLESTRAID: *South Wales*: bridge end, 22
  gallons.

LLESTRAID:
  Cardiff: of corn, 20 gallons = 2½ bushels =
    4 peccaid = 16 pedwran or quarters.
  Cowbridge: 22 gallons.
  Neath and *Swansea*: 22 or 24 gallons, the
    latter called a stacca.
LOAD:
  of bulrushes, 63 bundles. 27 G. 3.
  of hay, 36 trusses of 56 pounds each.
    11 G. 1.
  of lead, 'carrus,' '120 stone or 1,500
    pounds;' sometimes 168 stone, or 12
    wegs. In the Peak lead, sometimes 30
    fotmals, of 70 pounds each, making
    2,100 pounds or 8 score and 15 stone,
    that is, of 12 pounds each; of wheat,
    'summus,' the same quantity. 31 Ed. 1.
    But a load of wheat is 5 quarters.
  of wood, 50 cubic feet. 11 G. 1.
  of earth or gravel, 1 cubic yard = 27
    cubic feet.
  of lime, 32 bushels.
  of oak bark, 45 cwt.
  of timber, round, 50 cubic feet; square
    40 cubic feet.
  of sand, 36 bushels.
  of Scotch coal, 1 cwt.
  of lead, sometimes 175 pounds.
  Many of the English counties had their
  peculiar customary loads, varying with the
  commodity.
  Wiltshire: of timber, for the navy, 50
    cubic feet.
  South Wales: of lime, about 18 barrels, or
    540 gallons.
  Jersey: of firewood, about 20 cubic feet.
  Scotland, in some places: of oatmeal,
    20 stone, Dutch, 25 English.
  of meal, 2 bolls of 16 pecks; sometimes
    a peck over.
  of coal, in some places 2 cwt.
  of coal, 12 cwt.
  of coal, a waggon load is sometimes
    16 cubic feet.
  Dublin: of hay, 4 cwt., or more commonly
    4½.
LUG or LUGG: *Dorsetshire*: of land, 15 feet
  and an inch; called also goad, used
  instead of a pole of 16½.
  Herefordshire: of coppice wood, 49 square
  yards.

Lug or Lugg:
Hertfordshire: 20 feet.
Wiltshire: a pole or rod of 15, 16½, or 18 feet.

Maen: *South Wales*: of wool, formerly ¼ cwt. = 4 topstens; latterly only 26 pounds.
Maise: *South Wales*: of herrings, 30 score of 21 each = 630.
Mark: formerly ¾ of a pound.
of French copper, 8 ounces avoirdupois. 12 C. 2.
of gold or silver thread, ½ of a pound.
at *Guernsey* and *Jersey*, the marc of Rouen = 8¾ ounces avoirdupois nearly.
Mast: of amber, or of cullen gold and silver, 2½ pounds. 12 C. 2.
Math: *Herefordshire*: mowing; a day's math is about an acre, or a day's work for a mower.
Measure: *Cheshire*:
of barley, 38 quarts = 9½ gallons.
of malt, 32 or 36 quarts = 8 or 9 gallons.
Westmorland: of oatmeal, 16 quarts.
Kincardineshire:
of English coal, 48 Scotch pints.
of lime, 64 pints.
Guernsey and *Jersey*:
of apples, about 8 Winchester bushels.
of potatoes, 14 pots = 7 gallons.
Merk: *Shetland*: of land, from ½ an acre to 2 acres.
Mile: 8 furlongs = 320 poles = 1,760 yards = 5,208 feet. 35 Eliz.
Wales: Milldir, formerly 1,000 ridges of land, each of 3 leaps, or 20¼ f. making near 4 miles English.
Scotland: Nairn and Moray, in the cross roads, the old mile of 2,640 E. yards, nearly a mile and a half.
Mount: of plaster of *Paris*, 3,000 weight, 12 C. 2.
Mutchkin: *Scotland*: ¼ pint = ½ a chopin = 4 gills = 26⅓ cubic inches.

Nail: of cloth, $\frac{1}{16}$ yard = 2¼ inches.

Ounce: avoirdupois, $\frac{1}{16}$ pound = $\frac{7000}{16}$ = 437½ grains Troy.
Troy, 480 grains; formerly 452 for money.

Ounce:
of electuaries or drugs, 20 pennies. 31 Ed. 1 = 480 grains; the ounce of the apothecaries, the same as the Troy ounce.
Scotland: 476 grains, $\frac{1}{16}$ of a Dutch or Troy pound.
Ox-Land: *Glamorganshire* and *Pembrokeshire*: 8 customary acres.

Pace: commonly 2½ feet; the Roman pace was near 5 feet.
Wales: formerly 2¼ feet = 3 feet of 9 inches.
Pack:
of yarn, 4 cwt., each of 120 pounds.
of teazles, 9,000 heads of kings; 20,000 of middlings.
Gloucestershire: of teazles, 40 staffs = 1,000 glens = 20,000.
of kings, 30 staffs = 900 glens = 9,000.
Huntingdonshire: of wool, 240 pounds.
Kent: of flax, 240 pounds.
Yorkshire, N. R., of teazles, 1,350 bunches of 10 each = 13,500.
North Wales: of lambs' wool, 240 pounds; but of Yorkshire and Lancashire lambs' wool, 44 pounds.
Clydesdale, *Dumfries-shire*, and *Selkirkshire*: of wool, 12 stone Sc.
Packet: of leaf metal, 250 leaves. 11 G. 1.
Paladr: *Anglesey*: the perch of 4½ yards square = 20¼ square yards.
Palm: sometimes denotes 3 inches, as the Italian *palmo*.
Pared: *Montgomeryshire*: of cloth, 8 yards.
Peccaid: *South Wales*: of corn, 4 pedwran, of 5, 5½, or 6 quarts each, making 5, 5½, or 6 gallons; called also a hobaid.
Montgomeryshire: 5 gallons, called also a hoop.
Anglesey and *Carnarvon*. *See* Peget.
Peck: ¼ bushel = 2 gallons = 4 quarterns.
of flour and salt, generally reckoned 14 pounds.
Gloucestershire: of potatoes and green vegetables, about *Bristol*, 2 pecks striked; at *Gloucester*, a heaped peck.
Northumberland, *Alnwick*, and *Wooler*: ¾ of a Winchester bushel.
Newcastle: barley and oats, 5 forpits or quarterns.

PECK:

North Wales: of potatoes, 24 quarts.

South Wales: Llanbedr, 20 quarts.

SCOTLAND: $\frac{1}{4}$ firlot, nearly, $\frac{3}{8}$ Winchester bushel except for wheat.

of meal, 8 pounds Dutch, $8\frac{3}{4}$ English.

Aberdeen:

of ground malt, weighs from 12 to 14 pounds Dutch.

of potatoes, 32 pounds Dutch = 35 English.

Argyleshire: of potatoes; *Campbeltown*: of 9 wine gallons English heaped, weighing 56 pounds, avoirdupois. *Inverary*: 14 pints and a mutchkin, about $6\frac{1}{2}$ English wine gallons.

Banffshire: of potatoes, 2 strike = 32 pounds Dutch.

Berwickshire: $\frac{1}{3}$ of a firlot, instead of $\frac{1}{4}$.

Clydesdale:

of apples and pears, about 18 pints Sc. $= 6\frac{1}{2}$ Winchester gallons, called a sleck.

of meal, $\frac{1}{2}$ stone = 8 pounds Dutch.

Cunningham: of potatoes, formerly 32 pounds of 24 ounces each; now reduced to 27.

Dumbartonshire: of potatoes, the water peck nearly 42 pounds.

Kincardineshire: of potatoes, 2 stone Dutch.

Kintyre: of barley, bear, malt, and oats, a measure 12 inches in diameter, $10\frac{1}{10}$ deep, containing $1,142\frac{3}{10}$ cubic inches, a little more than half a Winchester bushel; formerly heaped, now striked.

Lanarkshire: of beans and pease, $\frac{1}{3}$ less than of barley.

Glasgow: of potatoes, 42 pounds avoirdupois.

Renfrewshire: of potatoes, from 36 to 37 pounds avoirdupois.

Sutherland: of potatoes, 28 pounds Dutch.

PECK LOAF: to weigh 17 pounds 6 ounces avoirdupois. 31 G. 2.

PEGET: *Anglesey* and *Caernarvon*:

of corn, 2 hobeds = 4 storeds = 16 kibins = 8 Winchester bushels, or a quarter.

of lime, 4 Winchester bushels.

PENNY: formerly $\frac{1}{240}$ of a money or Tower pound.

—

PENNYWEIGHT: at present 24 grains, $\frac{1}{240}$ of a pound Troy.

PERCH, POLE, or ROD: a measure of length, equal to $5\frac{1}{2}$ yards = $16\frac{1}{2}$ feet. 35 El. The same measure squared is employed as the first element of the acre, which contains 160 square perches of $30\frac{1}{4}$ square yards each. A cubic rod of $166\frac{3}{8}$ cubic yards is sometimes used. In many counties a perch of 8 yards is used for fencing. The forest pole is 7 yards; in Sherwood Forest, 25 feet. A coppice pole is 6 yards.

PERCH: *Berkshire*: sometimes 18 feet for rough work.

Devonshire:

of stone work, $16\frac{1}{2}$ feet in length, 1 in height, and 22 inches in thickness.

of cob work, 18 feet in length, 1 in height, and 2 in thickness.

Herefordshire: of fencing, 7 yards in length; of walling, $5\frac{1}{2}$.

Hertfordshire: sometimes 20 feet, sometimes called a lug.

Lancashire: $5\frac{1}{2}$, 6, $6\frac{1}{2}$, 7, $7\frac{1}{2}$, or 8 yards, in different parts of the county.

Leicestershire: of hedging, 8 yards; sometimes 8 yards square for land.

Oxfordshire; of draining, 6 yards.

Westmorland: near *Lancashire*, 7 yards.

Guernsey: 7 yards squared for land measure, making $1\frac{5}{8}$ perches.

Jersey: $7\frac{1}{3}$ yards = 22 feet, $\frac{1}{90}$ of an acre.

South Wales: of land.

1. Sometimes 9 feet square, 160 making 1 stangell; 4 stangells 1 erw of 5,760 square yards.
2. Sometimes $10\frac{1}{2}$ feet square, called a quart, or quarter of a llath, 40 of which make a stangell, whence the erw is 7,840 yards, equal to the Irish acre.
3. Sometimes 11 feet, called bat or eglwys haw, make an erw of 9,384 yards; or in *Glamorganshire* $\frac{1}{5}$ more = 11,261, reckoning 48 to the rood or quarter stang.
4. Sometimes $11\frac{1}{4}$ feet, called a llath, 48 making a quarter cyvar, and 4 cyvars an erw of 11,776 yards.
5. Sometimes 12 feet, called a quart or quarter llath, giving an erw of 10,240

PERCH:
South Wales:
yards, equal to the *Staffordshire* acre.
of labourers' work; in some parts of Wales, 6, 7, or 8 yards.
Scotland: 18¾ feet.
Dumfries-shire: a rod of 3 ells, or 9 feet 3 inches.
Ireland: of land, 7 yards in length or square.

PIECE:
of calico, 10 yards or less. 4 G. 3.
of linen, cambrics, or French lawns, 13 ells. 38 G. 3.
of sailcloth, 38 yards, ¾ wide. 1 Ja. 1.
Derbyshire: of lead, at the cupolas, or smelting houses, 176½ pounds.
Dorsetshire:
of flannel, 85 yards, yard wide.
of sailcloth, about 40 yards, yard wide.
Hampshire: of calico, 28 yards, ¾ wide.
Northampton: of stuffs, 32 yards, 22 inches wide.
Shropshire: of flannel, 100 yards.
Dumfries-shire: of carpet, 63 yards.

PIG: of lead, 21½ stone = 301 pounds.
Derbyshire: at the smelting house, 352½ pounds.
Northumberland: 1½ cwt. = 168 pounds.

PINT: ⅛ gallon = ½ a quart = 28⅞ cubic inches, wine measure; 35¼ customary ale measure, 33$\frac{6}{10}$ Winchester measure.
Scotland: 2 chopins, about 105 cubic inches, 3 ale pints E. But the standard jug, which was entrusted in 1621 to the care of the magistrates of Stirling, appears to contain only 103¾ cubic inches, or 103$\frac{7}{10}$. The standard at Aberdeen contains 105$\frac{6}{10}$ cubic inches. The ale pint of Stirling is $\frac{1}{16}$ above the standard.
Argyleshire: the customary pint contains 109·87 cubic inches.
Caithness: the standard pint contains 18 gills, for ale, or ⅛ above the regular standard; but it has a pin within, marking 16 gills, which is used for measuring spirits.

PINT WEIGHT: of butter, in Norfolk and Suffolk, 1¼ pound.

PIPE: of wine, ½ a tun = 2 hogsheads = 126 gallons, otherwise a butt. 2 H. 6; 5 Anne. More commonly, however, of Lisbon, 140 gallons; of Port, 138; of Sherry, 130; of Mountain, 126; of Vidonia, 120; of Madeira, 110.
Guernsey and *Jersey*: of cider, 240 pots, about 120 gallons.

PLOUGHLAND: *Wales*: 8 oxlands = 64 customary acres.

POCKET: of wool, ½ a pack = 120 pounds.
of hops, *Kent*, 1½ cwt.
Surrey, 1¼ cwt, measuring about 5¾ feet in circumference, 7½ long; 4 pounds being allowed for the weight of the canvas.

POKE: of wool, 20 cwt.

POLE. *See* Perch.

POT:
of ale, generally a quart.
of butter, 14 pounds. 13 & 14 C. 2.
Guernsey and *Jersey*, about 2 quarts.

POTTLE: 2 quarts.

POUND: *avoirdupois*: 16 ounces, 7,000 grains. *See* Ounce.
Troy: 12 ounces of 480 grains each = 5,760 grains.
of money and drugs formerly contained 20 shillings; of all other things, 25. 31 Ed. 1.
of Venetian gold, 12 ounces Venice weight. 12 C. 2; said to be about ½ an ounce more than the 'pound mark.'

POUND:
of husks or nuts, 21 ounces. 12 C. 2.
of raw silk, 24 ounces. 24 G. 3.
of silver coins, a pound sterling; the money pound, or Tower pound of the Anglo-Saxons, used for some centuries after the Conquest, was $\frac{15}{16}$ of our Troy pound. In various parts of the country a pound contained 12, 16, 17, 18, 20, 22, 24 ounces.
Guernsey and *Jersey*: a little more than 17 ounces; the same of bread, 1½ pound, avoirdupois.
North Wales, 18 to 21 ounces.
South Wales, 17, 18, and 24.
Westmorland, 12, 16, 18, and 21 ounces.
Scotland: Trone pound, 1¼ pound Dutch

POUND:
  Scotland:
    or Scotch Troy = 20 ounces, 21¾ ounces avoirdupois.
    Troy or Dutch weight, 16 ounces = 7,621 7/10 grains E. = 1·0888 pounds avoirdupois = 17·42 ounces avoirdupois; in Amsterdam, it appears to be reduced to 7,000 grains.
  Aberdeenshire:
    of butter and cheese, 20 or 26 ounces Dutch.
    of malt, meal, meal and corn, 24 ounces Dutch.
  Angus: Trone pound, 22 ounces avoirdupois, in some places 24, 26, or 27 ounces.
PUNCHEON:
  of beer, in London, 72 gallons.
  of wine, 84 gallons.
PWYS: *South Wales*: of wool, about 2 pounds, 1/12 of a maen.

QUART: 2 pints, whether of wine measure or ale measure; of dry measure, a peck is sometimes called a quart.
  Scotland: two Scotch pints.
  WEIGHT, *Gloucestershire* and *Leicestershire*: of butter, 3 pounds.
  Isle of Man: of wool, 7 pounds, ¼ of a quarter.
  Measure of length or surface, Wales: a pole of 3½, 4, or 4½ yards.
QUARTER: 8 bushels.
  of salt, 4 cwt.
  Devonshire: of Welsh coal or culm, 16 heaped bushels.
  Derbyshire: of lime, at the wharfs, 8 level bushels; at the kilns, 8 heaped bushels.
  Yorkshire:
    of chopped bark, in some parts, 9 heaped bushels.
    of oats, for bread, in some parts to be made up 3 cwt. See Bushel.
  Guernsey and *Jersey*: of potatoes, 240 pounds Dutch weight = 263 avoirdupois.
  South Wales: of corn, 2½ quarts, for quartern, or quarter peck.
  Llanbeder: 2 pecks, ½ a bushel.
  Banffshire: 8 bushels and 3 pecks, Winchester measure.

QUARTER:
  Guernsey and *Jersey*:
    of apples, 4 measures = about 12 Winchester bushels.
    of land: *South Wales*, ¼ cyvar, 40 perches. See Perch.
    WEIGHT, ¼ cwt. in *Guernsey* and *Jersey*: 25 pounds of Rouen, 27⅛ avoirdupois.
QUINTAL: properly 100 pounds; sometimes written kintal.
  of cheese, in some counties, 120 pounds.
  of line, 25 bushels.
QUIRE: of paper, 24 sheets.

REAM: of paper, 20 quires.
REEL: *Scotland*: 90 inches. 13 C.1. Used for yarn in *Clydesdale*.
  Essex: short reel for wool, 1¼ yard; long reel, 1½.
  Hampshire: for flax, 2 yards round.
REES: of herrings, 15 glences = 375.
RHAW: of peat, *Wales*, 140 or 120 cubic yards, 280 square yards 18 inches deep, or 15 poles of 4 yards square each, making 240 square yards, of the same depth.
RIDGE: of land, *Wales*, formerly 21¼ feet, or 3 leaps.
ROD. See Perch.
ROLE: of parchment, 72 sheets. 12 C. 2.
ROOD: of land, properly ¼ acre = 40 perches = 1,210 square yards; but the term is often provincially used for rod, or a measure approaching to it.
  Cheshire:
    of hedging, 8 yards.
    of land, 8 yards square = 64 square yards.
    of marl, 64 cubic yards.
  Cumberland: 7 yards.
  Derbyshire:
    of bark, seems to be a pile 7 yards in length.
    of draining or fencing, 7 or 8 yards.
    of digging, 7 yards square.
    of slating, 9 square yards, 44 square yards.
  Durham: of wall building, 7 yards.
  Northumberland: 7 yards.
  Shropshire:
    of hedging, 8 yards.
    of digging, 8 yards square.

ROOD:
Warwickshire; of fencing, sometimes a perch, or 5½ yards.
Westmorland; of slating, 6½ yards square = 42¼ square yards.
Yorkshire: in the moorlands, of fencing, 7 yards.
Wales: of ditching, draining, and hedging, 8 yards.
Berwickshire:
of labourer's work, 6 or 7 yards.
of masonry, 6 yards square, 2 foot thick.
Dumbartonshire: 6 yards square.
Dumfries-shire: of draining, 19 foot = 6 yards 1 foot.
Fifeshire: of fencing, 6 yards.
W. Lothian: of draining, 6 yards.
ROPE: in some places, 20 feet.
Devonshire: of cob-work and masonry, 20 feet in length, 1 foot high, and 18 inches thick.
Somersetshire: of wall building, 20 feet in length.
ROUL: of minsters or ozenbrigs, 1,500 ells. 12 C. 2.
RUCK: of bark, in some parts, 5¾ cubic yards, stacked.
RUNDLET or RUNLET: of wine, 18 gallons.

SACK:
of coal, 3 bushels; the sacks to be 50 inches by 26. 3 G. 2.
of flour or meal, 280 pounds. 37, 39, 40, 54 G. 3.
of meal, 5 bushels. 6 G. 1; 2½ cwt. = 280 pounds. 31 G. 2.
of salt, 5 bushels.
of wool, 2 weys, or 26 stone = 3¼ cwt. 14 Ed. 3; 25 Ed. 3; 11 H. 7.
Gainsborough: 2 bushels, to be 52 inches long, 26 wide. 9 G. 8.
Bedfordshire: of corn, 5 bushels.
Devonshire: of coals, in some parts, 1¼ cwt.
Dorsetshire: of flour and grain, 4 bushels of 9 gallons each = 4½ Winchester bushels.
Essex: of charcoal, 8 pecks.
Gloucestershire: of potatoes, 3 bushels, or 2½ cwt.
Hertfordshire: of ashes, 5 bushels; 4 striked, 1 heaped.

SACK:
Kent: of apples and potatoes, about 3½ bushels.
Somersetshire: of potatoes, 240 pounds.
Surrey:
of charcoal, 5 bushels.
of oats, 4 bushels.
of potatoes, 3 bushels of 60 pounds each.
Warwickshire: of beans and wheat, 3 bushels of 9 gallons each.
Wiltshire:
of beans, peas, wheat, and vetches, usually 4 bushels.
of potatoes, 36 gallons, or 2 cwt.
Worcestershire: of apples, 4 bushels.
Yorkshire, W.R.: of potatoes, 14 pecks = 3½ bushels.
North Wales: of wheat, 1½ hobaid, to weigh 260 pounds.
SAUME: of quicksilver, 315 pounds meaning probably somme, a load.
SCORE: Derbyshire: of lime, 20 or 22 heaped bushels.
Liverpool: of barley, beans, and oats, 21 bushels.
Newcastle: of chaldrons of coals, 21.
North Wales: of hobeds, 21.
Dumbartonshire: of sheep, sometimes 21.
Roxburghshire and Selkirkshire:
of bolls of grain, sometimes 21.
of sheep, sometimes 21, adding 1 more to every hundred.
SCRUPLE: apothecaries' weight, 20 grains.
SEAM or SEEM: sometimes a quarter of corn or malt.
of glass, 3 cwt., or according to 31 Ed. 1, 6 cwt., 28 stone of 24 pounds each.
Devonshire: of dung, 3 cwt.
SESTER: of wheat, before the Conquest, was a horse load. Hoveden, in Tyrrel.
SHILLING: of silver coin, before the Conquest, 1/40, afterwards 1/20 pound.
SHOCK: of canes or boxes, 60. 12 C. 2.
Derbyshire: seems to mean 12 sheaves of corn.
SIEVE: Kent:
of apples and potatoes, about a bushel.
of cherries, 48 pounds.
SKAIN or SKEIN: Hampshire: of yarn, 480 yards.

SKAIN or SKEIN:
 Suffolk: of yarn reeled, 20 leas, each of 80'
  or 120 yards, making 1,600 or 2,400.
SPAN: 9 inches.
SPINDLE: Clydesdale: 48 cuts of 120 threads
  each, on a reel 2½ yards round, making
  48 × 300 = 14,400 yards.
SQUARE: frequently 100 square feet.
 Derbyshire: of slater's work, 100 square
  feet.
 Devonshire:
  of thatching, 10 feet square = 100 square
   feet.
  of timber, 6 inches; as a mode of expressing the quantity by the length only,
   it is reduced to the supposed size of 6
   inches square.
STACCA: South Wales: sometimes a llestraid
  of 3 Winchester bushels.
STACK: Bedfordshire: of hard wood, 1 yard
  square and 4 in length = 4 cubic yards.
 Derbyshire: of coals, the three-quarter
  stack in some parts is 105 cubic feet,
  2 less than 4 cubic yards.
 Middlesex: of wood, 4 cubic yards.
 Northamptonshire: of fire-wood, 4 cubic
  yards.
 Shropshire: of coals, a cube of 4 feet,
  weighing about 25 cwt.
STAFF: of teazles; Essex: 51 bunches, or
  gleans of 25 each = 1,250.
 Gloucestershire: 25 glens of 20 = 500; of
  kings 30 glens of 10 = 300.
STANG or STANGELL: South Wales: ¼ crw.
  See Erw, Perch.
STICK: of eels, 25. 31 Ed. 1.
STONE: formerly in London, 12½ pounds.
  31 Ed. 1; that is, ⅛ of 100 pounds, instead of ⅛ of 1 cwt.
 of alum, 13½ pounds.
 of glass, 5 pounds.
 of hemp, 20 pounds. 21 H. 8; sometimes 32 pounds.
 of hemp or flax, 16 pounds. 24 G. 2.
 of lead, 15 pounds, each 25 shillings in
  weight. 31 Ed. 1.
 of meat, 8 pounds.
 of wool, 14 pounds. 14 Ed. 3.
 Bedfordshire:
  of butchers' meat in the south, 8 pounds.
  north of the Ouse, 11 pounds.

STONE:
 Buckinghamshire: of cattle, 8 pounds.
 Cumberland: of hay, tallow, wool, or yarn,
  and sometimes of meat, 16 pounds.
 Durham: of wool, 18 pounds.
 Essex: of beef, 8 pounds.
 Kent: of meat, in some places, 8 pounds.
 Gloucestershire:
  of beef, 8 pounds.
  of wool, 12½ pounds.
 Herefordshire: 12 pounds.
 Liverpool: 20 pounds.
 Middlesex: of meat, 8 pounds.
 Northumberland: of wool, 24 or 18 pounds.
 Suffolk: of hemp, 14½ pounds.
 Sussex: of meat, 8 pounds.
 Westmorland: 14, 16, or 20 pounds.
  of butter, 16 pounds of 20 ounces each =
   20 pounds.
 Yorkshire: of wool, 16 pounds, ¼ more
  being allowed for the draught or turn
  of the scale.
 Western Moorlands: 17 pounds and ½
  more for the draught.
 Eastern Moorlands: 17 to 10 pounds.
 About Darlington: 18 pounds.
  of wheat, W.R., 22 pounds.
 North Wales: of wool, from 4 pounds to
  15 pounds.
 South Wales: of wool, 14 pounds with 1
  pound ingrain, making 15 pounds when
  sold to woolstaplers.
 In various markets, provincial weights
  of 4, 5, 6, 7, 11, 13, 14, 15, 17, 18, 21,
  22, 24, and 26 pounds.
  of butchers' meat, commonly 8 or 12
   pounds.
 Scotland: Dutch or French Troy stone, of
  16 pounds, and 8 stone nearly equal to
  10 stone English; 8½ stone Dutch are
  much more accurately 10½ English.
  Established by an Act of 1681; varying from 15 to 28 pounds all over the
  country.
 Dublin:
  of rough tallow, 15 pounds avoirdupois.
  of wool, 16 pounds.
STOOK: Devonshire: of threshers' work, 10
  sheaves, from 7 to 10 inches through
  at the band; sometimes 12 sheaves
  make a stook.

STOKED: *North Wales*: of corn, 2 bushels, or 4 kibins.
STRIKE: a measure of corn, varying in its contents from ½ to 1, 2, and 4 bushels.
Montgomeryshire: ½ a hobed of Denbighshire = 2 hoops = 40 quarts = 1¼ Winchester bushel. *See* Hoop, Bushel.
Fishguard: ½ a Winchester bushel. *See* Vat.
SUM: of nails, 10,000. 12 C. 2. Probably *somme*, a burden.

TABLE: of glass, 5 square feet.
TALSHIDE: a billet of certain dimensions, 4 feet in length, 'besides the carfe.' Number 1, 16 inches round; number 2, 23 inches. 34, 35 H. 8. If half round, to measure 19 inches about, instead of 16, and 18½ if a quarter round. 43 Eliz. These dimensions are made 10½, 12¼, and 12 inches respectively. 9 Anne.
TANKARD: of ale, a quart.
TEAL, TÈL, or TELAID: *South Wales*: *Llandovery*, 4 bushels of 10 gallons each, making 5 Winchester bushels.
Llanbedr: 4 quarters, each of 2 pecks, each of 5 gallons, making again 5 Winchester bushels.
Cardiganshire: in some parts 3 Winchester bushels; in others 6, called tèl mawr and tèl bach.
Brecknockshire and *Caermarthenshire*: of lime, in some parts, 4 bushels of 10 gallons each = 5 Winchester bushels equal to the Irish barrel used for coal and salt.
Pembrokeshire: 4 or 5 bushels, called a barrel; a long teal contains 8 Winchester bushels.
TERTIAN: of wine, 84 gallons. 2 H. 6.
THRAVE: of corn, *Derbyshire*: 2 kivers or shocks, or 24 sheaves.
of straw, *Gloucestershire*: 24 boltings or trusses, of 24 pounds each = 576 pounds. *See* Threave.
THREAD: of yarn; the circumference of a reel, *Suffolk*: 2 yards, sometimes 3; *Clydesdale*, 2½.
THREAVE: of straw for thatching; *Westmorland*: 24 sheaves.

THREAVE:
Yorkshire, E. R.: 12 bundles, not precisely limited in magnitude.
of corn in reaping, *Kincardineshire*: 2 stooks of 12 sheaves each, the sheaves at the band to fill a fork 10 inches wide.
TIERCE: of wine, 42 gallons, otherwise an aume.
TIMBER: of furs, 40 skins. 12 C. 2.
TOD: of wool, 2 stone = 28 pounds. 12 C. 2.
Bedfordshire: 28 pounds, and sometimes a pound over for pitchmarks, making 29 pounds.
Gloucestershire: 28¼ pounds.
Sussex: 32 pounds.
Yorkshire: Holderness, 28½ pounds.
Guernsey and *Jersey*: 32 pounds.
TON: 20 cwt. = 2,240 pounds avoirdupois.
of earth or gravel, a cubic yard is often reckoned a ton.
of wheat, 20 bushels.
On a canal, 40 cubic feet of oak or ash is to be considered as a ton. 23 G. 3. Sometimes 48 are allowed. In a ship's measure, 40 cubic feet are considered as a ton, being supposed to carry 2,000 pounds.
of barley, sometimes 1,700 pounds.
of lead, 19½ cwt.
of linseed oil, 236 gallons.
of oil, at the Custom house, 252 gallons.
of salt, 42 bushels.
Derbyshire:
of bark, gypsum, and lime, in some parts 2,400 pounds.
of coal on canals, 2,400 pounds.
of dung, hay, lead ore, stone, and straw, sometimes 2,400 pounds.
of grindstones, 15 cubic feet.
of broken stones, about Bakewell, 20 striked bushels.
Devonshire: of timber, 40 cubic feet.
Dorsetshire: of Portland stone, 16 cubic feet.
Essex: of potatoes, 2,520 pounds, or 36 bushels.
Leicestershire: of limestone, 5 quarters = 40 bushels.
Middlesex: Stratford, of sifted gravel, 23 cubic feet.

TON:
  Wiltshire: of timber, 40 cubic feet.
  Worcestershire: of coke, 28 bushels.
  Isle of Man: of coal and soapers' waste for manure, 48 bushels.
  North Wales:
    of coal on the sea coast, ⅔ chaldron, weighing 24 cwt.
    of lime, ½ of the ton of coals, or 12 cwt. = 16 Winchester bushels.
  South Wales: of culm, about 17 cwt.
  Berwickshire: of potatoes, shipped for London, 28 cwt.
  Clydesdale: of hay, nearly 100 stone.
  Kincardineshire: of potatoes, 4 bolls = 1 ton English
  Stirlingshire: of oak bark, 128 stone Dutch = 160 English, nearly.
TOPSTON: South Wales: of wool, 8⅓ pwys = 7 pounds = ¼ maen; latterly 6⅔ pounds.
TRUSS: of hay, 56 pounds if old; 60 pounds if new. 36 G. 3.
  Bristol: 7 pounds. 22 G. 2.
  London: formerly 36 pounds. 31 G. 2.
TUB:
  of butter, 84 pounds. 36 G. 3. In South Wales from 60 to 120 pounds, used as a measure for corn to be exported, 4 bushels. 2 G. 2.
  of tea, 60 pounds.
  of coal, see Corf.
TUN;
  of wine, 2 pipes = 252 gallons. 2 H. 6, 5 Anne; supposed by some to have been originally a ton weight, or rather 2,000 pounds, about 32 cubic feet, which is about 240 gallons of water, or about 252 Irish gallons.
  of beer in London, 2 butts.
  of oil, 252 gallons; of sweet oil, 236 gallons.
  of syrup, 3½ barrels.
TUNNELL: Cardiganshire: of lime, 16 bushels, about the produce of a ton of limestone.

VAT or FAT:
  of coals, ¼ chaldron = 9 bushels. 47 Geo. 3; sometimes called a strike.
  of corn, 8 bushels.

VERGÉE: Guernsey and Jersey: of land, 40 perches; a little less than ¼ an acre. See Perch.
VRAGINA: 2½ pennyweights. 31 Ed. 1.

WAIN: of coals, 17½ cwt. 6 and 7 W. 3.
WARP: of bed-ticking, Fordingbridge: 75 yards, from 2,000 to 3,000 threads.
WATER MEASURE: on board of ship, 5 pecks. 11 H. 7.
WAY or WEN: of glass, 60 bunches. 12 C. 2.
WED: of Ticking, Fordingbridge: 70 yards.
WEIGH or WEY:
  of cheese, flax, lead, tallow, and wool, 14 stone, 31 Ed. 1., properly 5 chaldrons, or 40 bushels.
  of cheese, 2 cwt.; but in Essex, 256 pounds. 9 H. 6. Otherwise 416, and in Suffolk, 3 cwt.
  of meal, 48 bushels, of 84 pounds each. 27 G. 3.
  of salt, 1 ton = 40 bushels. 38 G. 3.
  of window glass, 60 cases.
  of wool, 13 stone = 182 pounds.
  Devonshire: of lime, coal or culm, sometimes 48 double Winchester bushels.
  Dorsetshire: of wool, a weigh or weight is 30 pounds, and ⅓ pound or 1 pound over in some places.
  South Wales: of coals, 6 chaldrons = 8 tons 2 cwt,
Swansea:
  of refuse coal, about 9½ tons.
  of culm, 10 tons, or 216 heaped bushels.
WEIGHT: Dorsetshire: of hemp, 8 heads of 4 pounds twisted and tied, making 32 pounds.
Somersetshire: of hemp, 30 pounds.
WINDLE: of corn, Lancashire: 3 bushels; of barley, beans, and wheat, 3½ Winchester bushels.
  of straw, Mid-Lothian, $\frac{1}{20}$ kemple = 5 or 6 pounds trone weight.

YARD: 3 feet = 36 inches; but by custom, the legal yard for cloth has become 37 inches in many cases.

YARD:
Buckinghamshire:
  of land, from 28 to 40 acres.
  of bark, sometimes 37 inches.
Wiltshire: ¼ acre.
Isle of Man: of cloth, 38 inches.
Anglesey and *Carnarvon*: 40 inches.

YARD:
Anglesey and *Carnarvon*: of hay, 38 inches in length, of various height and breadth, the bargain being often made for a certain stack, according to its length only.

III.

A BILL TO AMEND THE LAW RELATING TO MEASURES OF LENGTH, AREA, BULK, WEIGHT, AND VALUE

BE it enacted by the Queen's most Excellent Majesty, by and with the advice and consent of the Lords Spiritual and Temporal, and Commons, in this present Parliament assembled, and by the authority of the same, as follows:

1. This Act may be cited as the Measures Act, 1895.
2. This Act shall not come into operation until the first day of January one thousand eight hundred and ninety-six, which day is hereinafter referred to as the commencement of this Act.
3. The same measures shall be used throughout the United Kingdom.
4. The bronze bar and the platinum weight, more particularly described in the first part of the First Schedule to this Act, and at the passing of this Act deposited in the Standards Department of the Board of Trade, shall be the imperial standards of measure of length and of weight, and the said bronze bar shall be the imperial standard for determining the imperial standard mete for the United Kingdom, and the said platinum weight shall be the imperial standard for determining the imperial standard yasp for the United Kingdom.
5. The four copies of the imperial standards of measure and weight, described in the second part of the First Schedule to this Act, and deposited as therein mentioned, shall be deemed to be parliamentary copies of the said imperial standards.

The Board of Trade shall as soon as may be after the commencement of this Act cause an accurate copy of the imperial standard of length measure and an accurate copy of the imperial standard of weight measure to be made of the same form and material as the said standards, and it shall be lawful for

Her Majesty in Council, on the representation of the Board of Trade, to approve the copies so made, and the copies when so approved shall be of the same effect as the said parliamentary copies, and are in this Act included under the name parliamentary copies of the imperial standards of length measure and weight measure.

6. If at any time either of the imperial standards is lost or in any manner destroyed, defaced, or otherwise injured, the Board of Trade may cause the same to be restored by reference to or adoption of any of the parliamentary copies of that standard, or of such of them as may remain available for that purpose.

7. If at any time any of the parliamentary copies of either of the imperial standards is lost or in any manner destroyed, defaced, or otherwise injured, the Board of Trade may cause the same to be restored by reference either to the corresponding imperial standard, or to one of the other parliamentary copies of that standard.

8. The secondary standards of measure and weight which, having been derived from the imperial standards, are at the commencement of this Act in use under the direction of the Board of Trade, and are mentioned in the Second Schedule to this Act, shall be secondary standards of measure and weight, and shall be called Board of Trade standards.

If at any time any of such standards is lost or in any manner destroyed, defaced, or otherwise injured, the Board of Trade may cause the same to be restored by reference either to one of the imperial standards or to one of the parliamentary copies of those standards.

The Board of Trade shall from time to time cause such new denominations of standards, being either equivalent to or multiples or aliquot parts of the imperial measures ascertained by this Act, as appear to them to be required, in addition to those mentioned in the Second Schedule to this Act, to be made and duly verified, and those new denominations of standards when approved by Her Majesty in Council shall be Board of Trade standards in like manner as if they were mentioned in the said schedule.

It shall be lawful for Her Majesty by Order in Council to declare that a Board of Trade standard for the time being of any denomination, whether mentioned in the said schedule or approved by Order in Council, shall cease to be such a standard.

9. The standards of measure and weight which are at the commencement of this Act legally in use by inspectors of measures for the purpose of verification or inspection, and all copies of the Board of Trade standards which after the commencement of this Act are compared with those standards and verified by the Board of Trade for the purpose of being used by inspectors of measures

under this Act as standards for the verification or inspection of measures, shall be called local standards.

10. The straight line or distance between the centres of the two gold plugs or pins (as mentioned in the First Schedule to this Act) in the bronze bar by this Act declared to be the imperial standard for determining the imperial standard mete measured when the bar is at the temperature of melting ice and when it is supported on bronze rollers placed under it in such manner as best to avoid flexure of the bar, and to facilitate its free expansion and contraction from variations of temperature, shall be the legal standard measure of length, and shall be called the imperial standard mete, and shall be the only unit or standard measure of extension from which all other measures of length, area, or bulk shall be ascertained.

11. One tenth part of the imperial standard mete shall be a hand, one tenth of such hand shall be a quil, and one tenth of such quil shall be a jot. Ten such metes shall be a beam, ten such beams a course, and ten such courses a reach.

12. A square mete shall be a nap, a hundred such naps shall be an ar, a hundred ars shall be a worth, and a hundred worths shall be an ing. The hundredth part of such nap shall be a foil, the hundredth part of such foil shall be a seal, and the hundredth part of such seal shall be an en.

13. The unit or standard measure of bulk from which all other measures of bulk, as well for liquids as for dry goods, shall be derived, shall be the litre or cubic hand. One thousand litres shall be a vat, and one thousand such vats shall be a keep. One thousandth part of such litre shall be a die, one thousandth part of such die shall be an ove.

14. The weight in vacuo of the platinum weight mentioned in the First Schedule to this Act, and by this Act declared to be the imperial standard for determining the imperial standard yasp, shall be the legal standard measure of weight, and shall be called the imperial standard yasp, and shall be the only unit or standard measure of weight from which all other measures of weight shall be ascertained.

15. One thousandth part of the imperial standard yasp shall be a gram, and one thousandth part of such gram shall be a une. One thousand such yasps shall be a ton, and one thousand such tons a poid.

16. Every contract, bargain, sale, or dealing, made or had in the United Kingdom for any work, goods, wares, or merchandise, or other thing which has been or is to be done, sold, delivered, carried, or agreed for by weight or other measure, shall be deemed to be made and had according to one of the imperial measures ascertained by this Act, or to some multiple or part thereof, and if not so made or had shall be void ; and all tolls and duties charged or collected

according to weight or measure shall be charged and collected according to one of the imperial measures ascertained by this Act, or to some multiple or part thereof.

17. The Act 41 and 42 Vict. Ch. 49, entitled Weights and Measures Act, 1878, is hereby repealed as to Sections 1 to 20 inclusive.

FIRST SCHEDULE

PART I. WILL DESCRIBE THE IMPERIAL STANDARD METE AND YARD CONSTRUCTED UNDER THE DIRECTION OF H.M. TREASURY; AND PART II. THE PARLIAMENTARY COPIES OF THE SAME.

Standards of the measures following are at the commencement of this Act in use under the direction of the Board of Trade.

Measures of Length

Beam: a chain of 100 links, each 2 hands long.
Half-beam: a chain of 50 links, each 2 hands long.
Double-mete, or fathom.
Mete: hand, quil, jot.
Mete divided into hands, quils, and jots.

Measures of Area

Nap divided into foils, seals, and ens.

Measures of Bulk

Vat: a cubic brass tank or cistern, 1 mete inside measure.
Half-vat: tank of same area, but half the height.
Cask: a cylindrical vessel containing 100 litres.
Half-cask: a cylindrical vessel half height of former.
Double-peck: brass cylinder containing 20 litres water.
Peck: cylinder containing 10 litres.
Half-peck: cylinder containing 5 litres.
Double-litre: rectangular vessel double height of litre.
Litre: a cubic vessel, 1 hand each way internally.
Half-litre: vessel of same area as last, half height.
Double-gill: cylinder containing a fifth of a litre.
Gill: cylinder containing one-tenth of a litre.
Half-gill: cylinder of 50 dies.
Spoon: glass vessel containing 10 dies.
Die: glass vessel; also cubic brass vessel 1 quil cubed.

Measures of Weight

Sack: a tenth part of a ton.
Half-sack: 50 yasps.
Double-wey: brass weight of 20 yasps.
Wey: 10 yasps.
Half-wey: solid brass cylinder of 5 yasps.
Double-yasp: brass cylinder of 2 yasps.
Yasp: solid brass cylinder.
Half-yasp. .
200-gram weight.
100-gram, 50-gram, 20-gram, 10-gram, 5-gram, 2-gram.
Gram, and half-gram.
200-une, 100-une, 50-une, 20-une, 10-une, 5-une, 2-une, une, and half-une.

THIRD SCHEDULE

Tables showing the Values of the Old Measures in Terms of the New, and the Values of the New Measures in Terms of the Old.

Length

Jot .	. .	·0394 in.	
Quil .	. . .	·3937 ,,	
Hand	. . .	3·9371 ,,	
Mete .	. 1 yard	3·3708 ,,	
Beam	. 10 ,,	33·7079 ,,	nearly ½ chain.
Course	. 109 ,,	13·079 ,,	
Reach	. 1,093 ,,	22·79 ,,	

Area

En .	·000001196 sq. yds.	
Seal .	·0001196 ,,	
Foil .	·01196 ,,	
Nap .	1·196 ,,	
Ar .	119·6033 ,,	
Worth	11,960·3326 ,,	
Ing .	1,196,033·26 ,,	under ⅔ sq. mile.

Bulk

Ove .		·00018 pint	
Die .		·00176 ,,	
Litre .	.	1·76077 ,,	equals ·11 peck.
Vat .	. 220 gal.	0·77 ,,	
Keep .	220,096 ,,		

Weight

Une .	. .	·00056438 dram	
Gram .	. .	·56438 ,,	
Yasp .	. .	2 lb. 3 oz. 4·3830 ,,	
Ton .	. 19 cwt. 76 ,,	9 ,, 15·04 ,,	nearly old Ton.
Poid 984 ton	4 ,, 13 ,,	4 ,, 5 ,,	

Value

Doit .	. .	·01 grams of gold	·3275 d.	
Groat .	. .	·1 ,,	3·275 ,,	
Cross .	. 1	,,	2s. 8·75 ,,	
Lion .	. 10	,,	1l. 7s. 3·5 ,,	

N.B.—Obviously the intermediate measures of bulk, weight, and value, if required, can be easily calculated from the above.

Imperial Measures in Terms of Basic System

Length

	Jots	R	C	B	M	H	Q	J
Inch	25·4 =						2	5·4
Foot	304·8 =					3	0	4·8
Yard	914·38 =					9	1	4·38
Fathom	1,829 =				1	8	2	9
Pole	5,029 =				5	0	2	9
Chain	20,116 =			2	0	1	1	6
Furlong	201,164 =		2	0	1	1	6	4
Mile	1,609,315 =	1	6	0	9	3	1	5

Area

	Ens	I	W	A	N	F	S	E
Sq. inch	645 =						6	45
Sq. foot	92,889 =					9	28	89
Sq. yard	836,097 =					83	60	97
Perch	25,291,989 =				25	29	19	89
Rood	1,011,677,600 =			10	11	67	76	
Acre	4,046,700,000 =			40	46	70		
Sq. mile	2,589,450,000,000 =	2	58	94	50			

Bulk, I.

	Oves	L		D	O
Cubic inch	16,386 =			16	386
Cubic foot	28,315,310 =	28		315	310
Cubic yard	764,510,000 =	764		510	

Bulk, II.

	Oves	L		D	O
Gill	141,980 =			141	980
Pint	567,930 =			567	930
Quart	1,135,870 =	1		135	870
Gallon	4,543,460 =	4		543	460
Peck	8,086,920 =	9		86	920
Bushel	36,347,700 =	86		347	700
Quarter	290,781,000 =	290		781	

Weight, I.

	Unes	T	Y	O	P
Dram	1,772 =			1	772
Ounce av.	28,349·5 =			28	349·5
Pound av.	453,590 =			453	590
Hundredweight	50,802,380 =		50	802	380
Ton	1,016,047,540 =	1	16	47	540

Weight, II.

	Unes			G	U
Grain	64·79895 =				64·79895
Pennyweight	1,555·17 =			1	555·17
Ounce Troy	31,103·5 =			31	103·5
Pound Troy	373,242 =			373	242

Value

	Dots	LN.	CR.	GT.	DT.
Penny	3·051 =				3·051
Shilling	36·612 =			3	6·612
Pound	732·238 =		7	3	2·238

www.ingramcontent.com/pod-product-compliance
Lightning Source LLC
Chambersburg PA
CBHW031954230426
43672CB00010B/2150